THE LIMITS OF SCIENCE

THE LIMITS OF SCIENCE

PETER MEDAWAR

OXFORD UNIVERSITY PRESS

Oxford University Press, Walton Street, Oxford OX2 6DP

Oxford New York Toronto
Delhi Bombay Calcutta Madras Karachi
Petaling Jaya Singapore Hong Kong Tokyo
Nairobi Dar es Salaam Cape Town
Melbourne Auckland

and associated companies in
Beirut Berlin Ibadan Nicosia

Oxford is a trade mark of Oxford University Press

First published in the USA 1984 by Harper & Row
as a Cornelia & Michael Bessie Book
First published in paperback in the United States 1987 by
Oxford University Press, Inc., 200 Madison Avenue,
New York, NY 10016, by arrangement with Harper & Row,
Publishers, Inc.

ISBN 0–19–505212–9 (pbk.)

Printed in Great Britain by
Richard Clay Ltd.
Bungay, Suffolk

TO JEAN

Contents

List of Illustrations

Preface

This is a serious book, but a very short one. I decided to make it so for two chief reasons: In the first place, I have long thought that nearly all books on nearly all subjects—especially the philosophical—are much too long. As a student at Oxford, I was dazzled, exhilarated, frightened and encouraged by witheringly cynical and sardonic philosophic tutorials with the Kantian philosopher Thomas Dewar (Harry) Weldon—tutorials arranged by my guide and mentor J. Z. Young as part of an intellectual upbringing such as only Oxford can provide—but I was nearly put off philosophy altogether by the extreme length, leaden prose and general air of joyless learning characteristic of the principal writings of Alfred North Whitehead. Another incentive to brevity was the fact that in later years the writings I found most exciting and eye-opening were very short—most of them no more than essays. Among them I include Sir Philip Sidney's *Apologie for Poetrie* (1580); Descartes' *Discours de la Méthode* (1637); Samuel Taylor Coleridge's *A Preliminary Treatise on Method* (1818), the first volume of the *Encyclopaedia Metropolitana;* Shelley's *Defence of Poetry* (1821) against the opinions expressed in "The Four Ages of Poetry" (1820)—another very short book—by his old friend Thomas Love Peacock. To these I add Immanuel Kant's *Introduction to Logic*— not really a book, or at any rate not a book by Kant; simply the title given by T. K. Abbott to his edition (1885) of a number of Kant's lectures on logic, put together after Kant's death by his pupil Jäsche (vol. 3 of the Rosenkranz *Sämmtliche Werke*). To these I unhesitatingly add A. J. Ayer's *Language, Truth, and Logic* (1936); nor must I forget two very short professional works that profoundly influenced many immunologists other than myself: Karl Landsteiner's *The Specificity of Serological Reactions* (1936) and *The Production of Antibodies* by F. M. Burnet and Frank Fenner (1949). These writings, of the shortness of which no one has complained, seemed to me to justify my resolution to be brief.

The three essays which make up this book are in three different styles. For "An Essay on Scians" I followed the aphoristic style adopted by Francis Bacon and William Whewell in a number of their writings. It is the style of exposition most considerate of the reader's interests, for the paragraph headings protect him very effectively from the discomfiture of learning what he does not wish to know, or of following discussions to the outcome of which he is completely indifferent.

The second essay, "Can Scientific Discovery Be Premeditated?" began as a lecture delivered on 5 June 1980 at a joint meeting of the American Philosophical Society of Philadelphia and the Royal Society of London, to which I am grateful for permission to reproduce it here.

The principal essay, "The Limits of Science," is written in the format of a short book. I am a professional scientist and a lover of science; there could be no greater misunderstanding of the purpose of this third essay than to suppose it to be in any sense antiscientific in mood. The purpose of "The Limits of Science" is simply to exculpate science from the reproach that science is quite unable to answer those ultimate questions I refer to repeatedly in the text—questions I show to be beyond the explanatory competence of science. But in spite of this failing, science is a great and glorious enterprise—the most successful, I argue (p. 65), that human beings have ever engaged in. To reproach it for its inability to answer all the questions we should like to put to it is no more sensible than to reproach a railway locomotive for not flying or, in general, not performing any other operation for which it was not designed. I have lectured upon the limits of science in several places and have received friendly criticisms, which I took note of and have introduced into the text where they deserved serious attention. A good deal turns on a "Law of Conservation of Information" (pp. 78–82)—a title suggested to me by an article written thirty or forty years ago by a Dr. H. A. Rowlands, titled "The Law of Conservation of Knowledge." I have failed completely in my attempts to find and reread this article.

My account in "The Limits of Science" of the significance for seventeenth-century science of the Pillars of Hercules was based upon that of Professor Marjorie Hope Nicolson (then of Smith College), and I am very grateful to Professor Sir Ernst Gombrich for calling my attention to more recent research which shows that the story is a good deal more complicated than she made it out to be. Indeed, Sir Ernst has repeatedly helped me by dipping into his vast storehouse of knowledge of cultural history. I have benefited also from criticisms by Professor Sir Karl Popper, all of which I have accepted, and for his corrections of certain Kantian usages I had not got quite right.

I take this opportunity to thank my wife and co-author, Jean, for proposing innumerable stylistic improvements throughout the text, and my secretary and assistant, Mrs. Joy Heys, for preparing the book for press.

AN ESSAY

ON

SCIANS

That which in the English-speaking world is known by everyone as "science" was not always so designated. The *Oxford English Dictionary* proffers the following homophones: sienz, ciens, cience, siens, syence, syense, scyence, scyense, scyens, scienc, sciens, scians. Of these, I like best the one I have chosen, though I am sad that the others are out of work. All derive, of course, from *scientia,* knowledge, but no one construes "science" merely as knowledge. It is thought of rather as knowledge hard won, in which we have much more confidence than we have in opinion, hearsay and belief. The word "science" itself is used as a general name for, on the one hand, the procedures of science—adventures of thought and stratagems of inquiry that go into the advancement of learning—and on the other hand, the substantive body of knowledge that is the outcome of this complex endeavor, though this latter is no mere pile of information: Science is *organized* knowledge, everyone agrees, and this organization goes much deeper than the pedagogic subdivision into the conventional "-ologies," each pigeonholed into lesser topics. Science is, or aspires to be, *deductively ordered:* It parades principles, laws and other general statements from which statements about ordinary particulars follow as theorems. The sciences don't begin this way, to be sure; nor do they always end in this tidy, deductively ordered form. Samuel Taylor Coleridge, writing *A Preliminary Treatise on Method*, complained that zoology as he studied it was so weighed down and crushed by a profusion of particular information, "without evincing the least promise of systematizing itself by any inward combination of its parts," that it was in danger of falling apart. It was of course Linnaeus and Darwin and those whom they inspired who rescued zoology from the odium of being no more than a heap of ostensibly unrelated facts. All the sciences that we judge mature have the kind of internal connectedness which Coleridge deplored the absence of in zoology. This kind of connectedness, or holding-togetherness, gives the sciences great stability and power to assimilate more information: In a long-established science it is impossible to imagine a situation in which a single phenomenon could so shake its foundations that it might all come tumbling down. Correspondingly, science is never long in a turmoil of self-questioning about its fundamental premises and assumptions. Another property that sets the genuine sciences apart from those that arrogate to themselves the title without really earning it is their predictive capability: Newton and cosmology generally are tested by every entry in a nautical almanac and corroborated every time the tide rises or recedes according to the book, as it is also corroborated by the periodic reappearance on schedule of, for example, Halley's comet (next due, 1986). I expect that its embarrassing infirmity of prediction has been the most important single factor that denies the coveted designation "science" to, for example, economics.

The Truth. Truth, we have learned, is of many different kinds, not all fully compatible. There is spiritual or religious truth, for example, and there is poetic truth—that which Sir Philip Sidney, in the spirit of Aristotle, defended against the coarse, hectoring notion of truth that went with being an historian (or a scientist).

In history and science the imagination was cramped and confined. by—in Sidney's own words—the "historian's bare *was*"; whereas the poet may represent nature or the past as it ought to be or ought to have been, that is, in a "doctrinable form"—a form meet for the teaching of salutary lessons.

Scientific truth in the sense I shall describe and explain it below is often thought of as the goal of a scientist's work, though "asymptote" would be the better word, for there can be no apodictic certainty in science, no finally conclusive certainty beyond the reach of criticism.[1]* There is no substantive goal; there is a direction only, that which leads toward ultima Thule, the asymptote of the scientist's endeavors, the "truth."

The scientist is in bondage to that conception of truth which has brought it about that aircraft fly and physicians sometimes cure their patients. It is the "correspondence theory of truth," which Alfred Tarski[2] has done so much to clarify. He put it thus: "A true sentence is one which states that a state of affairs is so and so and the state of affairs is indeed so and so." For example, the proposition "The atomic weight of potassium is 39" is true if and only if the atomic weight of potassium indeed is 39. Such a declaration always excites a derisive bray from philistines who see in it still further evidence of the philosophers' preoccupation with trivialities. To say or think so is, however, to miss the point. What Tarski is in effect saying is that the notions of truth and falsity are metalinguistic concepts, for it is only sentences or "propositions" of which the truth or falsity can be affirmed

---

*Notes will be found at the end of each essay.

or denied. Any such affirmation or denial must therefore be a statement about a statement, delivered in that language—metalanguage—in which we discourse about language. Like Moses in the battle against the Amalekites, metalanguage has two arms which must at all times be kept aloft: logical syntax and semantics. Logical syntax deals with the rules of reasoning (of deduction, for example) and of sentence transformation generally, and semantics has to do with the notions of meaning and of truth. This is a humbly commonsensical conception of truth but it evidently works, for science is built upon it. It is not a conception that lights up the fine arts or indeed has much relevance to them except under special circumstances, such as, for example, the false attribution of the authorship of a painting, sonata, or poem.

The most heinous offense a scientist as a scientist can commit is to declare to be true that which is not so; if a scientist cannot interpret the phenomenon he is studying, it is a binding obligation upon him to make it possible for another to do so. If a scientist is suspected of falsifying or inventing evidence to promote his material interests or to corroborate a pet hypothesis, he is relegated to a kind of half-world separated from real life by a curtain of disbelief; for as with other human affairs, science can only proceed on a basis of confidence, so that scientists do not suspect each other of dishonesty or sharp practice, and believe each other unless there is very good reason to do otherwise.

Unintelligibility of Science; Specialization. The degree to which science is cut off from humanity generally by its reputed incomprehensibility is easy to overestimate. People who *want* to understand the gist of science do so; Barbara Ward, admittedly an extremely intelligent woman—but not a scientist—did so, and her book *Only One Earth: The Care and Maintenance of a Small Planet* (written with René Dubos, 1972) is a monument to what intelligence and the will to learn can do to the barriers thought to isolate science from the world. In any case, the *ideas* of science are often quite simple. Such an idea as "the mass of the earth" is not at all difficult to grasp. What *can* be difficult to grasp is the means by which its value is ascertained: it is the scientific performance rather than the scientific conception that tends to bewilder the lay public.

As to the specialization of scientists: Commencement addresses and other public utterances about nothing (appropriately enough often entrusted to prominent citizens with nothing to say)harp interminably upon how the increasing specialization of scientists makes them unintelligible to each other and a fortiori to the lay public. This generalization about scientists, as spelled out elsewhere (pp. 71–72), is simply not true: Scientists are getting less specialized, not more. This revelation is likely to cause some awkward gaps in commencement addresses, gaps that might with profit be filled up with something original, or at least true.

Science and the Mandarins' Fingernails. It is said that in ancient China the mandarins allowed their fingernails—or anyhow one of them—to grow so extremely long as manifestly to unfit them for any manual activity, thus making it perfectly clear to all that they were creatures too refined and elevated ever to engage in such employments. It is a gesture that cannot but appeal to the English, who surpass all other nations in snobbishness; our fastidious distaste for the applied sciences and for trade has played a large part in bringing England to the position in the world which she occupies today. It was not always so. The people who today look down on and sneer at applied scientists would in Charles II's day have been poking fun at the virtuosi of the Royal Society, many of whom were engaged in "pure" science of what was widely thought to be a specially ludicrous kind: weighing air, for example—can you beat it? Whatever next? The playwright Thomas Shadwell (1642?–1692) caught the mood exactly in his *The Virtuoso*. When the curtain goes up, Sir Nicholas Gimcrack is seen making froglike swimming movements on the table in his workroom. Does he intend to swim in the water? he is asked, and he replies: Never, sir; I hate the water. "I content myself with the speculative part of swimming and care not for the practical." And: "I seldom bring anything to use; it is not my way. Knowledge is my ultimate end." As the play unfolds, other equally ludicrous ambitions are discovered. Thus there was tension between "pure" and "applied" science then as there is now. Of those who spoke of this tension, Bacon's is, as always, the voice that carries farthest, particularly when his imagery makes it so easy to see why he was long reputed the author of Shakespeare's plays. "Experiments of use," he said in the Preface to *Novum Organum* (1620), were not enough; there must also be "experiments of light." The light was Bacon's own special light, *lumen siccum,* the light of understanding.

Thomas Sprat, in his great history of the Royal Society (1667), upheld Bacon's distinction, adding that to complain that the findings of science did not lead immediately to the results of practical usefulness was as silly as to complain that not all seasons of the year were seasons of harvest and vintage; but in spite of this advocacy, the superiority of pure to applied science came to achieve the status of an axiom—something so self-evidently true as to require no demonstration. It was not only that applied science was essentially banausic in character—fit only for execution by menials or slaves—but also that the axioms or principles of the pure sciences, such as theology, their reputed queen, were made known not by any vulgar empirical exertions such as observing or measuring things or (ugh!) experimenting with them, but by intuition or revelation, and were thus—as the empirical sciences never can be—apodictically certain.

Scientists. Scientists have been known by as many names as the *scians* they profess, and Charles Onions in his *Dictionary of English Etymology* (Oxford, 1966) cites *sciencer, sciencist, scientman* and *scientiate.* The old-fashioned term was "man of science," meaning what modern manners would oblige us to call a science person. All such terms were superseded by the proposal of the greatest ever nomenclator, William Whewell, Master of Trinity College, Cambridge (pp. 14, 33, 51, 84), who, in his introduction to *The Philosophy of the Inductive Sciences* (1840), wrote: "We very much need a name to describe a cultivator of science in general. I propose to call him a 'scientist'."

Mathematicians are reputed to be rare and special people, exulting in the exercise of a gift far beyond the performance, and perhaps even the conception, of ordinary people. Scientists are not; quite ordinary people can be good at science. To say this is not to depreciate science but to appreciate ordinary people. But to be good at science one must *want* to be—and must feel a first stirring of that sense of disquiet at lack of comprehension that is one of a scientist's few secure distinguishing marks. I think it is the lack of this exploratory, hunting trait that makes it unthinkable to many people who could be scientists that they should be so. Good scientists often possess old-fashioned virtues of the kind schoolteachers have always professed to despair of ever inculcating in us. These are: a sanguine temperament that expects to be able to solve a problem; power of application and the kind of fortitude that keeps scientists erect in the face of much that might otherwise cast them down; and above all, persistence, a refusal bordering upon obstinacy to give up and admit defeat.

Science and Cricket. To dispel at once any gloomy forebodings such a paragraph heading may give rise to, let me say at once: No, it is not possible to project a cricket ball by design in such a way that upon touching the ground it first breaks (i.e., takes a sharp turn) in one direction and at the second bounce breaks the other way. The two ideas were conjoined in my mind because I once read a convincing article by a senior West Indian on the social function of cricket in the British West Indies. The writer declared that high proficiency in cricket opened for a youngster a casement giving onto a world he might never otherwise have known: a prosperous and enjoyable livelihood, a chance to meet new people and to enjoy unlimited opportunities for travel. It gave, moreover, a certain place in life, a measure of self-esteem and a reason for feeling it.

This is all parable, but I use it to emphasize that all these things may also come the way of a schoolboy or schoolgirl who may be quite ordinary in the sense of not having been born into a wealthy or specially well-educated family or having enjoyed the real or supposed advantages of a costly private education. For such a one, science can do what cricket is reputed to do for West Indian youngsters. Because a scientific career does not call for rare, high or unusual capabilities, it is accessible to almost all, and a career in science stands out as one of the great opportunities of a liberal and democratic society. Moreover, science itself is various enough to satisfy all temperaments. "Among scientists," I once wrote, "are collectors, classifiers and compulsive tidiers-up; many are detectives by temperament and many are explorers; some are artists and others artisans. There are poet-scientists and philosopher-scientists and even a few mystics . . . and most people who are in fact scientists could easily have been something else instead."

Science and Culture. One of the ways—understandable but gravely diminishing—in which the English have traditionally revenged themselves upon the Americans for being so prosperous and generally good at many of the things we would have liked to be good at ourselves is by despising them for being such upstarts: so nouveaux, don't you know. This whole attitude is epitomized by the snobbish story in which an Oxbridge college gardener tells a brash American visitor how he, too, could have just such a perfect lawn: "Just roll it and water it and roll it and water it for three or four hundred years." I shall scream if I hear this story again.

It is in the same mean-minded spirit that our brethren of humane arts avenge themselves on scientists for being so busily and to all appearances so happily employed and for getting so big a cut of the governmental grant—for are they not the nouveaux riches of the campus, these half-educated and barely articulate tradesmen with coarse or unawakened sensibilities, with whom conversation at table is such an ordeal?

But come now. Is it not the function of a "university" worthy of the name to enculture (new word) these unhappy people?—something that could of course be done only by laying on special cultural lectures for scientists; for in any modern university it is taken for granted that nothing can be known, or perhaps even *should* be known, about subjects upon which students have not attended lectures. Mercifully, however, the whole hideous scheme of piping on cultural lectures for scientists came to nothing in any university I have been associated with, thus sparing some well-meaning mediocrity the embarrassment of watching the bored writhings of students of chemical engineering morally coerced into attending lectures on the English novel or on the origins of the Romantic movement in Germany. A young scientist who has not the initiative to read books, listen to music or visit art galleries, and argue about cultural likes and dislikes, is in a plight that cannot be remedied by cultural lectures which he will probably have the good sense not to attend. It would be a different story, of course, if there were an Ernst Gombrich or a Kenneth Clark on every campus, but people who can see the whole picture are in grievously short supply. What *can* be found, though, on every campus are libraries, radios, records and dozens of enthusiasts anxious to communicate and share their enthusiasms. I also take the view —which would certainly have been Dr. Samuel Johnson's—that students of humane letters are not in the least likely to benefit from attendance at lectures on "the" scientific method (whatever that may be) or on the supposed "principles" of physics, chemistry or genetics. The trouble is that young students of Eng. Lit. simply don't want to know much of what the average science lecturer would be able to impart to him. Long experience has taught me that people nagged at by an anxiety to improve the minds of their fellow men are in real life among those least likely to be able to do so.

An Apologie for Science? All who study science have their attention drawn to the obligation upon them to exculpate science from crimes of which it is not guilty: of causing a deterioration in the quality of life and dashing our hopes of its improving, of despoiling the environment and the like. I remember from many years ago a complaint in a literary weekly that the benefactions of medical science were so difficult to reconcile with the general proposition, the truth of which was taken for granted, that science works everywhere for the diminishment of man.

A former editor of the London *Sunday Express* (a newspaper with a very big circulation) once wrote: "Science gave us the Great War," committing thus the elementary blunder of blaming the weapon for the crime, though not even he, I imagine, would have gone so far as to blame science for nationalism, bungling politicians and ambitious generals such as those who at Passchendaele and on the Somme so nearly did in the British army during the First World War.

Science is blamed as a matter of course for the barren, blackened soil of the industrial midlands in England and of the eastern seaboard of the United States as we may see it on the railroad journey from New York City to Philadelphia; it is blamed for the despoliation of the countryside and for pollution too, though these are almost wholly the work of nineteenth-century laissez-faire capitalism and the philosophy of commercial advantage, something for which the interests of the environment are a notorious impediment—environmentalists are a confounded nuisance, no less. The blaming of science is a movement of thought from which scientists get no comfort. They must learn to regard as a principal social function of science to act as scapegoat for blunders and malefactions of its political masters. So far from taking this conciliatory view, however, scientists are more likely to feel as I do, aggrieved at science's being incriminated for ravages of the environment brought about by the collusion of negligence, self-interest and greed.

Scientific Inference. Sciential reasoning, wrote Samuel Taylor Coleridge in his *Aids to Reflection* (1825), "is the faculty of concluding universal and necessary truths from particular and contingent appearances." This direction of flow of thought is described as "inductive," and it was long assumed that induction was the method characteristic of science.

Although later he so far questioned this belief as to wonder whether the word should not be abandoned altogether, William Whewell certainly did at one time believe in induction. At the very beginning of his *History of the Inductive Sciences* (1837), he wrote: "The advance of science consists in collecting general laws from particular facts and combining several laws into one higher generalisation in which they still retain their former truth." The principal advocates of induction were John Stuart Mill (1806–1873) in his *A System of Logic* (1843) and Karl Pearson (1857–1936) in his *The Grammar of Science* (1892); and its principal opponents, William Whewell (1794–1866) in his *Philosophy of the Inductive Sciences* (1840) and Karl Popper in his *The Logic of Scientific Discovery* (1959). I have gone pretty thoroughly into the rights and wrongs of induction in *Pluto's Republic,* and need not go over the same ground again. The essential point is that there is no logically rigorous procedure by which an inductive "truth" can be proved to be so. Even in the simplest kind of iterative induction, such as philosophers are wont to illustrate by, for example, "All swans are white," all we know for certain is, in this case, that this swan is white and so is that and the other, but the transition from these particulars to the audacious generalization that *all* swans are white has no ratiocinative justification; it involves an act of mind—or act of faith—that Whewell called "superinduction," for such a generalization cannot contain more information than the sum of its known instances. To suppose otherwise would be to flout the "Law of Conservation of Information" (pp. 78–82). Candidate laws of supposedly inductive origin have all kinds of disabilities.

They are prey to a number of grave paradoxes that offend logic or common sense. Thus Bacon saw that a simple iterative induction can be upset by a single contradictory instance, something that does not happen to scientific hypotheses in general.

Of the paradoxes, I shall mention only two that afflict these simple enumerative inductions. Our sense of the fitness of things is offended by the idea that an induction such as "All swans are white" can be corroborated by the discovery in a trash heap of an old black boot, yet so it is: for if all swans are white, it follows logically that all non-white objects are non-swans. If, then, any black object is discovered which anxious scrutiny shows not to be a swan, then we have confirmed a logical prediction from a hypothesis and given ourselves an extra incentive to believe in it. The skeptical philosopher Sextus Empiricus, who lived in the third century A.D., propounded a strictly logical paradox that impressed John Stuart Mill and, in his *Studies and Exercises in Formal Logic* (1884), John Neville Keynes (father of John Maynard). Here it is:

Consider a syllogism of which the major premise is an inductive generalization:

> *Major premise:* All men are mortal.
> *Minor premise:* Socrates is a man.
> *Conclusion:* Socrates is mortal.

The charge brought against this syllogism by Sextus Empiricus was that of *petitio principii*—begging the question—for unless we *already* know that Socrates is mortal, how can we be so sure "all men are mortal," as the major premise declares?

There are two ways out of this ostensibly grave objection: (1) The syllogism should be read not in the asseverative but in the conditional form: if . . . , and if . . . ; then . . . (2) There is just the same element of *petitio principii* in every example of deductive reasoning, for, as explained on pp. 79–80, all that deduction can do is to bring out into the open something already present in the premises. Indeed, all Euclid's theorems are question begging. None of them says anything more than is contained in his axioms and postulates.

For these and other reasons, I go along with the opinion of William Whewell, Bertrand Russell and Karl Popper that scientists do not make their discoveries by induction or by the practice of any other one method. "The" scientific method is therefore illusory. "An art of discovery is not possible," said William Whewell, and more than a century later we can say with equal confidence that there is no such thing as a calculus of discovery or a schedule of rules by following which we are conducted to a truth. Why, then, did so clever a man as John Stuart Mill believe that there could be propounded rules such as he outlines in Chapter 8 of his *A System of Logic?* One reason was that he did not really understand the exploratory character of science. He writes as if he believed that the scientist would have already before him a neatly ordered pile of information ready-made—and to these he might quite often be able to apply his rules, as some kind of "machine intelligence" (p. 81) might too. The procedure was so skillfully lampooned by Lord Macaulay that it is no longer taken seriously. Moreover, Mill did not draw a clear distinction in his mind between the methodologies of discovery and of proof.

I give now an example of scientific inquiry from everyday life—one that illustrates the comparative ordinariness of the talents and simplicity of the logic needed, and makes it clear why residence on the foothills of Mount Parnassus is not necessary. A busy housewife finds that the lamp on her worktable isn't working. Why not? It can't be a fuse or the wall plug that's wrong, because she can still use the iron off it; the bulb, moreover, works in the lamp by the sewing machine. The switch is of unsound design and has given trouble before—can it be that? She decides to take the whole contraption over to a plug which is known to be in order; it doesn't work there, so that is probably it. It won't take a moment to fit on that spare switch and then lo, the lamp goes on again. The scenario I have just outlined is that of the "hypothetico-deductive" method as William Whewell envisaged it, and differs only in degree of difficulty from that which a scientist employs in studying more difficult and more important problems: The logical procedures are straightforward and anyone can use them who has the wits to stay alive in a complex modern world offering us so many easy opportunities to depart from it. It is characteristic of this procedure that the empirical information assembled by observation and experiment is that which is required by the formulation of the hypothesis itself: No information is collected in vacuo.

But why do scientists sometimes like to think—as Darwin certainly did—that they proceed by induction? It can only be because the myth of induction is that which accords best with the self-image a scientist may have formed of himself: as a regular, straightforward, plain-thinking man of facts and calculations—someone very different from a philosopher, a poet fellow or an imaginative writer. He is a veritable Thomas Gradgrind, as Dickens pictured him in *Hard Times* (1854).

The truth is that there is no such thing as "scientific inference." A scientist commands a dozen different stratagems of inquiry in his approximation to the truth, and of course he has his way of going about things and more or less of the quality often described as "professionalism"—an address that includes an ability to get on with things, abetted by a sanguine expectation of success and that ability to *imagine* what the truth might be which Shelley (p. 52) believed to be cognate with a poet's imagination.

Science and Politics. Some of my colleagues will think me sensible, others disloyal, when I roundly declare that political and administrative problems are not in general scientific in character, so that a scientific education or a successful research career do not equip one to solve them. Moreover, I do not believe that the coming of the millennium would not be long delayed if members of Parliament and congressmen were all to be trained in science. More than that, I regard my opinions as so obviously right as not to be worth the exertion of justifying them. It is only when the problems that confront us turn upon scientific evidence that one might hope for a larger measure of scientific understanding than currently prevails. In such situations as these, scientists should be consulted—and when their advice is concordant, it should be taken and acted upon. When I was director of the National Institute for Medical Research in London, the mayor of a large American city wrote me a civil letter asking my opinion of the fluoridation of water. I accordingly put before him the epidemiological evidence, and to help him appreciate the direction in which the evidence tended, I told him that every time an American municipality determined against fluoridation there was a little clamor of rejoicing in the corner of Mount Olympus presided over by Gaptooth, the God of Dental Decay. Of course, the more difficult part of the fluoridation enterprise is not scientific in nature—I mean that of convincing disaffected minorities that the purpose of the proposal is not to poison the populace in the interests of a foreign power or to promote the interests of a local chemical manufacturing company, a big employer of labor.

Many people who ought to know better derive their conception of science from boys' books or the gothic extravagances of science fiction. Soon after the Second World War, an Anglican bishop wrote to *The Times* of London, asking whether the Allies might not agree to destroy the formula of the atomic bomb as if it were a cooking recipe. A scientist's mind boggles at the magnitude of such a misconception.

Although politicians tend to have a low opinion and an unaccountably resentful attitude toward science and scientists, which I shall illustrate later, I believe it a general rule for their guidance (paraphrasing Bertrand Russell) that when experts are unanimous in holding a particular opinion, the contrary view cannot be regarded as certain. I have never met a scientist who does not believe that the effects of ionizing radiations from a nuclear bomb, especially upon deoxyribonucleic acid, the vector of genetic information, are so utterly frightful as to outweigh all other considerations in framing national policy, among which I include national pride (the chap inside is past caring whose flag flies over his tomb). Because of its degree of centralization, England is especially vulnerable to a nuclear war, and when I read of our civil defense proposals I feel that legislators underestimate by several orders of magnitude the gravity of the threat that confronts us. The havoc of nuclear warfare is not something that British grit and the Dunkirk spirit will see us through, and the hardships that confront the civil population amount to more than the milkman's missing his round and that dratted newsboy's being late again with the morning paper. Both may well have been vaporized.

I recount now the best single example known to me of synergy between science and government. During the Second World War, the necessity for rationing in Britain put it within the power of central government to determine not only how much people ate but also the nature of what was eaten. The then Minister of Food, Lord Woolton, sought scientific advice on nutritional science—and took it. Physiological research had already made known the essential ingredients of a diet suitable for youngsters, and this diet was now made available to them, by law, with the happy consequence that the war babies of 1939–1945 grew up bigger and healthier than any previous generation in Britain. However, literary intellectuals made uneasy by what is ostensibly evidence of a benefaction of science can take comfort from the thought that this good wartime diet also accelerated the onset of sexual maturity, something which doubtless contributed to the reduction in the average age of marriage and may well have added to the proportion of extramarital births; but there—that's science for you.

No such happy story can be told of Edwin Chadwick (1800–1890), a culture hero of environmental science and the author of the hugely important *Report . . . on an enquiry into the sanitary condition of the labouring population of Great Britain* (1842), a work of the gravity and importance of *The Condition of the Working Class in England in 1844* by Friedrich Engels (1845). Chadwick's advocacy led to the passage of the Public Health Act of 1848.

Because Chadwick's proposals (like environmental legislation in England and America today) could put costly obligations upon public funds and on manufacturers, the act aroused a great deal of public displeasure. *The Times* appeared to think that dying in one's own way without governmental interference was an elementary democratic privilege, for in an editorial on the proposal to lay a main sewer in London, *The Times* commented: "We prefer to take our chance of cholera and the rest than to be bullied into health. England wants to be clean, but not cleaned by Chadwick."[4]

Among those who took his chance, as *The Times* preferred, was Queen Victoria's husband, Albert, the Prince Consort, who died of typhoid at age forty-two. The water in Windsor Castle was doubtless contaminated by the fifty-two overflowing cesspools found there after his death.

An example of the antiscientist attitude of the House of Commons has been recounted by an authority on medical statuary, Professor M. McIntyre (*History of Medicine,* March/April 1980).

So great a benefactor of mankind as Edward Jenner might surely be expected to rate as a national hero, but the proposal to erect a statue to him in Trafalgar Square was scathingly denounced in the House: "Jenner had no business amongst the naval and military heroes of the country," and would indeed "pollute and desecrate the ground." Jenner was accordingly removed to the north side of Kensington Gardens, though Boston and Boulogne have built statues to honor those who merely introduced vaccination into America and France respectively. (See also "Postural Antiscientism," pp. 36–37.)

The Art of the Soluble. Borrowing a form of words from Bismarck—or was it Cavour?—I once remarked that if the art of politics is indeed the art of the possible, then the art of scientific research is surely the art of the soluble. Very felicitous, I thought—and what is more, very true. A problem shown to be soluble is a problem halfway solved. I shall illustrate this from my own research. I am reputed to be—and am often introduced before lectures as—the chap whose work made modern organ transplantation possible.

This is simply not true, for there is no direct causal connection between my research and the triumphs of modern organ transplantation. Organ transplantation was introduced into medicine by sanguine and adventurous surgeons in France, America and England, and the first reports of their attempts to get away with it were met by me and my kind with pursed lips and slow, grave shakings of the head, for, identical twins apart, there had at that time been no single report of successful transplantation between genetically unrelated human beings. Some barrier prohibited the transplantation of cells and tissues between different individuals, and at that time it was very far from obvious that the barrier could ever be surmounted. For one thing, it had been in existence for hundreds of millions of years, being already fully developed at the evolutionary stage represented today by modern bony fish, and thanks to the work of George Snell in Bar Harbor, Maine, and Peter Gorer at Guy's Hospital in London, it was coming to be realized that substances that arouse the graft rejection reaction—the transplantation "antigens"—were genetically programmed and thus as much part of a person's inborn constitution as his blood group and as little likely to be alterable.

Because it is interesting and, as Sir Philip Sidney would have said, "doctrinable," I shall recount in full the story of how the transplantation problem came to be shown to be soluble. The story began in 1958 at an international congress of genetics in Stockholm. There I met a bright and friendly New Zealander, Dr. Hugh Donald, a livestock geneticist and at the time head of the Animal Breeding Research Organisation in Edinburgh, an agency of the Agricultural Research Council. Donald was using cattle twins to try to sort out the relative contributions of heredity and upbringing to the character makeup of cattle— milk yields, conformation and the like. Such an analysis turns upon making a clear distinction between, on the one hand, identical twins, which have the same genetic makeup and which can be reared under different conditions, and on the other hand, fraternal twins, which except in age are no more alike than ordinary litter mates, for the differences between fraternal twins brought up under so far as possible identical conditions are mainly or wholly genetical in origin. The entire scheme conformed to the principles established by Sir Francis Galton about a century before; but Donald was worried about the accuracy of his ascertainment of the types of twins, whether identical or fraternal, because upon this the whole enterprise depended. There is no difficulty of principle, I assured him. Let tiny skin grafts be exchanged between the members of each pair of twins. If the grafts were accepted, the twins must surely be identical, but if not, merely fraternal. "I'll do it for you," I said, my mind doubtless unhinged by liquor or by the kind of mateyness that tends to prevail at international congresses.

Later on, much to my dismay, Hugh Donald reminded me of my promise, adding that the enterprise was entirely feasible because the Agricultural Research Council, his employer, had assembled all his cattle twin pairs at a research station within thirty or forty miles of Birmingham, where at that time I held the chair of zoology at the university. Moreover, I should be able to rely upon the help of two able young veterinary officers. I discussed the whole project with my close colleague Rupert Billingham, who had already greatly impressed me by adding to his native intelligence a naval know-how and a practical-mindedness that resulted from his period of service in the navy —which he had joined at a rather bad time for the Allies, with the purpose, which he accomplished, of bringing the war to a speedy and successful conclusion. He agreed that we should take on the job together.

The project met with two unexpected obstacles. One day on our drive to the farm where the twins were kept, we had a serious crash caused by a truck's pulling directly across our driveway (the truck's owner had frugally replaced a broken window in the driver's cabin with a rectangle of opaque sacking —an action to which the police drew the magistrates' attention in the prosecution that followed). Bill and I were quite badly bashed about, but lived to surmount the second of the two setbacks: My confident declaration in Stockholm of the certain success of the proposed procedure for distinguishing between the two kinds of twins by skin grafting now seemed likely to be falsified, for although twins which by all other criteria (blood groups, etc.) had been classified as identical duly accepted skin grafts from each other, so did twins that must certainly have been merely fraternal, being sometimes of different sex, as identical twins can never be. With our two veterinary helpers we went over the whole ground again, getting the same result: Fraternal twins would indeed accept skin grafts from each other.

We couldn't figure it out until we had read a, in many ways, remarkable book, *The Production of Antibodies* by F. M. Burnet and F. Fenner (1949). This book drew a parallel between antibody formation and the formation of adaptive enzymes in bacteria, and it contained the remarkable prediction that if that which would otherwise have been an immunity-provoking substance was presented to an organism early enough in life, the organism would not respond to it and would still not do so even if the immunity-provoking agent was presented when the organism became immunologically mature later on. What was important from our point of view was some of the evidence that Burnet and Fenner brought forward in favor of this unusual view, especially the remarkable findings of Dr. Ray Owen, who was at that time working with Dr. M. R. Irwin in the Department of Agricultural Genetics in Madison, Wisconsin.

What Owen had found was that fraternal cattle twins contained a mixture of red blood corpuscles: Each twin of a pair contained its own endogenous red blood corpuscles mixed in variable proportion with corpuscles belonging by right of genetic origin to its twin partner. This must surely have been a consequence of the fact that cattle twins, unusually, are synchorial—i.e., share one placenta—with the effect that the twins can transfuse each other with blood before birth though of course their individual blood circulations are quite separate from that of the mother. The exchange of red cells itself cannot have been enough: The twins must also have exchanged red cell precursors, for red cells went on being formed throughout life. Evidently the twins had performed the very experiment Burnet and Fenner's hypothesis invites one to try: They had confronted each other very early in life with their own distinctive antigens, which under normal circumstances would have caused graft rejection, and what Burnet predicted apparently happened, for because of the exchange of cells in fetal life the twins had become, as we put it, mutually "tolerant" of each other's cells and tissues. No wonder our dizygotic twins accepted skin grafts from each other, for they were already in a sense mixtures of each other's cells. They were already, in our terminology, "chimeras."

Our way ahead was now clear. We must reproduce by experimental means and at will the phenomenon that occurs as a natural accident in twin cattle. This problem was allocated to our student Leslie Brent as the subject of his Ph.D. thesis.

By hindsight, Billingham, Brent and I now realize that in executing this research program we had a wonderful stroke of luck, the nature of which we could not appreciate at the time because it was made possible only by our clumsiness and inexperience.

We inoculated into the fetuses of mice of one strain (CBA)

a number of tissue fragments and large clumps of cells coming from various organs of mice belonging to strain A, then, when the fetuses had grown up, we transplanted an A-strain skin graft upon the CBA mouse—and upon one such mouse the graft lasted more than five standard deviations longer than the mean survival time of such grafts on normal mice. Such a result would have been expected to occur merely by luck much less often than once in several thousand trials. Brent was awarded his Ph.D.

Where, then, did the luck come in? The little tissue fragments we handled were very difficult to work with and tended to clog up the superfine hypodermic needles we were obliged to use for injections into fetuses; the dose of cells was also very difficult to quantify. If we had been more experienced, we should have taken from the donor either white blood corpuscles or cells teased out from the spleen or lymph nodes.

We now know that if we *had* done this, we should not have discovered actively acquired tolerance, the reason being that the lymphoid cells found among leukocytes and in the lymph nodes and spleen are the cells that execute immunological responses. If, then, we had injected them in an attempt to induce tolerance, we should have killed the mice into which they were injected, for the grafts would in effect have attacked and rejected their hosts, causing the "graft-versus-host disease" that Billingham and Brent later discovered, interpreted and characterized.

This idea illustrates my later remarks about luck (p. 49): that to appraise the role of good luck in research, we must count against the discoveries that we attribute to good luck the discoveries we *don't* make through the intervention of bad luck—a number we cannot ascertain because the discoveries we don't make leave no trace.

The experiments on actively acquired tolerance I have just outlined had a profound moral effect upon the many surgeons and medical scientists who had been attracted by the exigencies of war wounds to the study of transplantation, because our experiments showed that, quite contrary to our earlier gloomy forebodings, the transplantation problem *was* soluble and that the supposedly insurmountable barrier that prohibited transplantation between animals of different genetic makeups could be overcome; therefore the attack upon it might be renewed with the virtual certainty of eventual success. Needless to say, experiments entailing the inoculation of cells into a fetus are clinically impracticable and indeed, in practice, organ grafts are transplanted nowadays under a protective screen of drugs that weaken the immune response. This is why I say that these early experiments did not lead directly to the successes of modern organ transplantation. Their effect was *moral* only—was only to inspirit the surgeons who finished the job by exorcising the bogey that the project was impossible in principle. The whole story illustrates the appositeness of describing the art of research as the "art of the soluble," as I did above.

I should like to revert now to my saying that we could so easily not have discovered tolerance through attempting to induce it by the inoculation into fetal mice of immunologically reactive cells because they are so easy to handle and so easy to quantify. In spite of their many advantages, such cells, injected, would have killed the fetuses by exposing them to an immunological attack against which they would be defenseless. Such a misadventure would have delayed the discovery of tolerance and the solubility of the homograft problem for quite a time. In science, though, where one fails another may succeed, and as it happened, a very bright immunologist, Milan Hašek, belonging to the great days before Czechoslovak science was virtually brought to a standstill by the Russian conquest and colonization of 1968, discovered tolerance independently of ourselves. Hašek's technique was to make a vascular bridge between two embryonated hens' eggs, the respiratory membranes of which he exposed by making windows in the eggs, then apposing and making a bridge of embryonic tissue between them. The embryos thus were brought into "parabiosis" with each other and after hatching turned out to have the same blood groups and to be unreactive toward each other's red cells (and also, as our team showed, later to skin grafts transplanted from one to the other)—an excellent example of the phenomenon of simultaneous discovery, very familiar to sociologists of science, who have completely falsified the idea that the uniqueness of discovery is the almost invariable rule. My colleagues and I had produced the same effect by withdrawing 20 milliliters of blood from one chicken embryo through one of the veins running in the respiratory membrane of the egg and then injecting that amount of blood into a second chicken embryo: This worked as well as Hašek's technique of parabiosis and was very much simpler and quicker. The point is that we would have got there anyway, no matter what blunders we might have made to begin with. It is noteworthy that the motives that prompted Milan Hašek to mount his experiments on parabiosis were entirely different from ours. I don't think Milan knew anything about Burnet's work at the time, but his ambition was by parabiosis to bring about vegetative hybridization along the lines advocated on theoretical grounds by Trofim Denisovich Lysenko.

There is no substance to Lysenko's view that such hybridization can induce genetic changes, and the fact that Milan Hašek's parabiotic experiment worked merely illustrates the familiar methodological truth that true inferences may sometimes be drawn from false premises.

Women in Science. The prowess of women in science has traditionally been the subject of what Sir Francis Bacon described as "peremptory fits of asseveration," to which I shall now add: There is no scientific or methodological reason why women should be worse than men at science—and no evidence that they are. Yet the exigencies of childbearing and the tradition of prior domestic obligations make women physiologically and perhaps also psychologically more vulnerable than men to influences that distract from scientific research. The question, then, of whether the performance—as opposed to the capability—of women in science is inferior to men's cannot be answered until the completion of a full-scale scientific (i.e., sociological) investigation of the matter. I have reason to believe that such an investigation is now in progress.

Men and women who revere Marie Curie for the scientific prowess that led to her winning the Nobel Prize on two occasions are apt to forget what is the most remarkable thing about her—that in spite of the intent and single-minded concentration that such a feat as she accomplished calls for, she raised a daughter, Irène, who, instead of, in the modern fashion, denouncing her parents and all their works and becoming a fashion model or dashing off to India to seek Enlightenment, became a Nobel Prize winner herself.

On Being Too Punctilious in the Execution of Experiments. It has been shrewdly observed that an experiment not worth doing is not worth doing well. After writing upon the prowess of women in science, I cannot forbear from recounting how an excess of punctilio on my part deprived my wife of the chance to make a quite important discovery. For about a year after graduation, the then Jean Taylor worked as a graduate student under Professor Howard W. Florey at the University of Oxford's School of Pathology, of which he was head. This was in those pre-penicillin days when his preoccupying interest was in the life history and function of white blood corpuscles of the kind known as lymphocytes—subjects upon which at that time we were almost wholly ignorant.

Florey had read a paper by a famous American pathologist, A. A. Maximov, who had declared that circulating blood lymphocytes might under some circumstances swell up and turn into large cells of the kind usually referred to as macrophages. Florey suspected that this was due to the contamination of his population of lymphocytes by macrophages themselves. Having now perfected the surgical technique for securing pure populations of lymphocytes by cannulation of a rabbit's thoracic duct, the principal lymph vessel in the body, Florey asked my wife-to-be to cultivate these cells and observe them intently by continuous photography through a microscope housed in a box at body temperature. At that time the conventional method of tissue culture was to grow the cells to be observed in a nutrient jelly prepared from the fluid portion of chickens' blood mixed with an extract of chicken embryonic tissue which caused the blood plasma to clot. When Jean told me of this, I expressed my horror at the idea of growing the lymphocytes of rabbits in a medium so foreign to them zoologically, and proposed instead that the lymphocytes be kept alive by growing them in the serum of rabbits. This she did. The lymphocytes were perfectly happy and moved around vigorously, as they always do. Florey was pleased and relieved to learn that they underwent no kind of transformation into macrophages. Now, Maximov had recommended the use of rabbits that had been sensitized by an infection with the parasitic roundworm Ascaris, and he also recommended that the culture medium be caused to contain a trace of what Maximov described as an "Ascaris extract." If my wife had followed Maximov's instructions to the letter in spite of my advice against doing so, she might very likely have witnessed the phenomenon described about twenty years later as antigen-induced lymphocyte transformation, the transformation being, as we now say, of ordinary lymphocytes into "blast cells" capable, as ordinary lymphocytes are not, of cell division.

The whole procedure advocated by Maximov sounded pretty absurd, not to say "unscientific." That was why I encouraged Jean to use culture media derived from the same species or preferably the same rabbit as that from which the lymphocytes for examination were removed—and to forget about the Ascaris nonsense, as I thought it to be; and thus, through my being too clever, an important opportunity was missed.

My advice to her embodied a twofold error of judgment, the first being that if one is repeating someone's work, one must repeat exactly what he did, however silly the procedure may seem to be to the experimenter. The second error of judgment was my failure to realize that many of the "classical" experiments performed by the nineteenth-century worthies were clearly done under conditions that would horrify us today in respect of lack of asepsis, the use of culture media not adequately buffered against variations of acidity or alkalinity, and nonobservance of other elements of what is nowadays thought of as good laboratory practice.

To look at the matter in another way—more, perhaps, in the spirit of Aesop—let me simply say that the moral of this story is: *It doesn't do to be too clever.*

Fraud in Science. Enough examples of fraud in science have been uncovered in recent years to have given rise to scary talk about "tips of icebergs" and to the ludicrous supposition that science is more often fraudulent than not—ludicrous because it would border upon the miraculous if such an enormously successful (p. 65) enterprise as science were in reality founded upon fictions.

In the opinion of the profession, dishonest fiddling of scientific data is to be classified as a minor neurosis, such as cheating at patience—something done to bolster up one's self-esteem.

The principal cause of fraud in science is a passionate conviction of the truth of some unpopular or unaccepted doctrine such as Lamarckism or what may be compendiously called the "IQ nonsense," a doctrine which one's scientific colleagues must somehow be shocked into believing.

The "self-monitoring" system that science is reputed to enjoy can be assumed to work very well in most cases, but there are special circumstances in which it breaks down. A case in point is the notorious frauds of Sir Cyril Burt, professor of psychology in University College London, who invented or manipulated data on the IQs of identical or fraternal twins brought up together or apart and presented them in such a way as to make an overwhelming case for the enormously preponderant influence of heredity (as opposed to upbringing) in the determination of IQs. Why were not these bare-faced frauds—which included the invention of colleagues—quickly shown up for what they were? The reason is that Burt told the IQ boys exactly what they wanted to hear, so they had no incentive to inquire deeply into the authenticity of his work. The Burt disclosures gave rise to the widespread suspicion that IQ psychologists generally were frauds; but this is not a charge that can legitimately be brought against any profession. Anyone who studied their writings deeply would incline to the opinion that their principal disability was not fraudulence but sheer stupidity. They are almost unteachable, too.

Guesswork in Science. The generative act in science is the proposal of hypotheses—that is, in making, as Whewell said, *guesses* (a usage that was also Bertrand Russell's). Many people, like John Stuart Mill, are offended by such a description. It is just as true to say that science proceeds by guesswork as it is to say that Mozart wrote many engaging tunes and Coleridge many ingenious rhymes. The trouble is only that these descriptions sound disrespectful in a context where something more like awed gravity is thought to be required.

Experiments on Animals. The paragraphs entitled "The Art of the Soluble" describe experiments carried out on animals: on cattle, on rats and mice, and sometimes on chickens and eggs. Unlike some of my colleagues, I believe the use of experimental animals to advance medical knowledge needs to be justified, because so many concerned people having nothing against science or scientists regard the use of experimental animals even for the benefit of human beings as a moral disfigurement of medical science.

Not even the most devout today are likely to take a stand on the theological opinion put by Thomas Love Peacock[5] into the mouth of the Reverend Dr. Gaster: ". . . nothing can be more obvious than that all animals were created solely and exclusively for the use of man." "How do you prove it?" said another guest. "It requires no proof," said Dr. Gaster, "it is a point of doctrine. It is written, therefore it is so."

Let us not forget, though, that human beings are animals too, and that their interests are no less worthy of attention than those of their poor relations. It is unthinkable that new drugs or medical treatments should be introduced into clinical use without some assurance that even if they are not efficacious they will at least not do harm. If these tests are not carried out upon laboratory animals, the likelihood is that they will be carried out upon human beings. Thus in his *Lettres Philosophiques,* written from England, Voltaire records with amused detachment rather than with any sense of outrage that when variolation to prevent serious attacks of smallpox was introduced into England from the Near East through the advocacy of Lady Mary Wortley Montagu, the procedure was tried out on condemned felons in Newgate jail, who were rewarded by being pardoned and by not getting smallpox. The intrinsically safer procedure of vaccination, introduced by Jenner, was introduced directly into clinical practice, though there can be little doubt that it was tried out first upon members of the lower classes.

Jenner's discovery is always made much of by those who are eager to prohibit the use of experimental animals altogether, because its introduction was founded upon clinical reasoning without laboratory experimentation. Why should this not apply even to greater innovations, such as the use of inactivated polio virus, as in the Salk procedure, to protect against poliomyelitis? The reason is straightforward enough: Salk vaccine could not have been devised unless Salk had been able to add to and draw upon the great body of knowledge about immunity that had been accumulated as a result of several generations of experiments upon animals.

I do not know any medical scientist who does not prefer in vitro—in effect, test tube—experiments to the use of animals to ascertain the potency and efficacy of drugs; but the ambition to substitute inanimate for animate systems for such purposes cannot be fulfilled when the phenomena we are studying pertain only to whole organisms and not at all to isolated cells: To have a headache or back pain or to contract multiple sclerosis is a privilege of whole organisms and it is these that must be studied if we seek the melioration of pain or of disease. To test for the ability of chemicals to cause cancer we can often test whether or not they cause mutations in microorganisms—but of course, whole animals had to be used to establish the correlation between the two properties in the first place.

Utopia and Arcadia. It accords very well with my conception (see below) of the long-term purpose of scientific inquiry that the age during which institutional science came into being was also above all others the age of Utopias: I think especially of Francis Bacon's *New Atlantis* (1626), Tommaso Campanella's *City of the Sun* (1602) and Andreae's *Christianopolis* (1619). The appeal to scientists of Utopias is not just that in them the scientist usually cuts a very good figure, but more that Utopias depend heavily upon a science-based technology about which the authors tend to be discreetly vague (there is talk in Campanella, for example, of wind power and of wheels within wheels); the appeal is that the Utopias embody the idea of improvement and melioration of just the kind that scientists think they have it in their power to promote or confer.

The old Utopias were contemporary societies come upon in far-off seas, but today it is space travel that has what was formerly the appeal of geographic exploration, so our Utopias of today are either far off in space or far off in time. Arcadia is the conception farthest removed from Utopia, for one of its principal virtues is to be pastoral, prescientific and pretechnological. In Arcadia, mankind lives in happiness, ignorance and innocence, free from the diseases and psychic disquiet that civilization brings with it—living indeed in that state of inner spiritual tranquillity which comes today only from having a substantial private income derived from trustee securities.

Postural Antiscientism. Mention has already been made (pp. 20–21) of the resentment aroused by Edwin Chadwick's proposal to endow London with a main sewerage system and of the contempt of Parliament for the proposal that Edward Jenner, the man who introduced vaccination against smallpox, should be allowed to rank among our national heroes. These, I think, were premonitions of the postural antiscientism that prevails widely today (by "postural" I mean the unquestioning, unthinking, almost reflexly contemptuous relegation to the devil of science and all its works and the attribution to it of all evils, especially those that are in reality due to political incompetence or commercial greed). I relate now an extreme example of postural antiscientism: in the *New York Review of Books,* 29 December 1966, the social critic and architectural writer Lewis Mumford quoted the following lurid passage from Edward McCurdy's edition of *Leonardo's Notebooks* (London, 1908, page 266), which describes the fearful and irresistible rampaging of a black-faced monster with bloodshot eyes and ghastly features—a monster against which the populace defended itself in vain. The passage runs:

> Oh, wretched folk, for you there avail not the impregnable fortresses nor the lofty walls of your cities nor the being together in great numbers nor your houses or palaces! There remained not any place unless it were the tiny holes and subterranean caves where after the manner of crabs and crickets and creatures like these you might find safety and a means of escape. Oh, how many wretched mothers and fathers were deprived of their children, how many unhappy women were deprived of their companions! In truth . . . I do not believe that

ever since the world was created has there been witnessed such lamentation and wailing of people accompanied by so great terror. In truth the human species is in such a plight that has need to envy every other race of creatures.

I know not what to do or say for everywhere I seem to find myself swimming with bent head within the mighty throat and remaining indistinguishable in death buried within the huge belly.

According to Lewis Mumford, this nightmarish passage presents the "reverse side of Leonardo's hopeful anticipations of the future": It is Leonardo's premonition of the havoc and despoliation that the advance of science and technology would wreak upon the earth and its inhabitants; but, he adds, "There is no way of proving this."

This being so, I proffer an alternative and entirely different interpretation. Leonardo is describing, essentially truthfully, what would have been known to him by hearsay: the virtually irresistible onward progress of a bubonic and hemorrhagic plague such as the Black Death, from the vector of which—the rat flea—hiding away or congregation for prayer would be equally inefficacious. Mumford's idea, I submit, was buncombe (p. 103, ch. 5, n. 2).

The Royal Society. The Royal Society of London for Improving Natural Knowledge—to give it its full title—is to represent in these pages the scientific academies that grew up during or not long after the revolution of thought that inaugurated institutional science in the seventeenth century. Founded in 1660 under the patronage of King Charles II, and still under royal patronage, enjoying the present fellowship of Queen Elizabeth and Prince Philip, the Royal Society is the oldest and most famous scientific society in the world. It is a private institution and not an agency of, though it is supported financially by, the government. Its motto is as simple and straightforward as the purpose it avows in its title. The motto, *Nullius in verba,* means "Don't take anybody's word for it"—least of all Aristotle's, they might have added, for they all thought so. Having it in mind that some of the earliest presidents of the Society were Christopher Wren, Samuel Pepys and Isaac Newton, we can hardly wonder that the seventeenth-century Czech theologian Jan Amos Comenius should have regarded it as the body that would realize his dream of instituting a pansophia—"a single and comprehensive scheme of human Omniscience" (i.e., all the things under Heaven that it has granted us to know, to say, or to do). Comenius accordingly showered the most splendid encomiums upon the Royal Society: "Blessings upon your heroic endeavours, illustrious Sirs!" he wrote. "I congratulate and applaud you and assure you of the applause of all mankind."

How, exactly, do the Royal Society and comparable bodies provide for the advancement of science? A clue to the answer is given by the fact that a large part of their private business provides for the candidature and election of potential members, for self-perpetuation is a principal function: Such a society must comprise membership of such a high degree of scientific proficiency that everyone regards it as a reward and a distinction to join its Fellowship. This in turn entails that the membership should be of just the right size—not as large as the Royal Society once was, so that election was hardly a distinction—nor so difficult to get into as to make candidates almost indifferent whether they succeed or not. The size, moreover, must be such that elections can be made from candidates while in their scientific primes instead of being deferred, as in some foreign academies, until later in life. The Royal Society scores well on all these points, but it is of course a weakness of all such bodies that not all who are members deserve to be and not all who fail to get in do not. Thus a candidate may not be proposed because friends are too indolent to put his name forward and try to raise support for him, or may be kept out through spite or the real or supposedly higher claims of an alternative candidate. Such faults are to be debited to human frailty and not to any short-comings of the scientific societies, which do by and large perform a useful function in the advancement of learning by the maintenance of high standards and through the authority in science that fellowship confers.

The Program and Purpose of Science. There is a famous passage in his *New Atlantis* in which Bacon describes the program of the new science as the *effecting all things possible,* a formula which some people regard as invigorating and inspiring and others as depressing and frightening, according to their temperaments; and later, in the fragmentary and little-known *Valerius terminus* (first published in 1734), he asks himself the true end of knowledge and says: "To speak plainly and clearly [Bacon seldom did otherwise], it is a discovery of all operations and possibilities of operations from immortality (if it were possible) to the meanest mechanical practice"—in science we are in fact to acquaint ourselves with the whole extent of our present and possible future estate.

As to the purpose of science, Bacon's name is so closely associated with the notion that its purpose is to secure power over nature ("human knowledge and power meet in one") that we forget that there are many more typical passages in Bacon in which he advocated a much more humbly meliorist position.

One can defend the view, which is also my own, that Bacon believed that the purpose of science is to make the world a better place to live in. In his preface to *The Great Instauration* (the compendious title of Bacon's entire system of philosophy), he wrote: "I would advise all in general that they would take into serious consideration the true and genuine ends of knowledge; that they seek it neither for pleasure, or contention, or contempt of others or for profit or fame, or for honour and promotion; or suchlike adulterate or inferior ends: but for the merit and emolument of life, and that they regulate and perfect the same in charity: for the desire of power was the fall of angels; the desire of knowledge the fall of man; but in charity there is no excess, neither men nor angels ever incurred danger by it." He then asks that we should take it as our aim "not to lay the foundation of a sect or Placit but of human profit and proficience." Moreover, in Book I of *De Augmentis Scientiarum* (1623), Bacon specifies that we are to make application of knowledge "to give ourselves repose and contentment and not distaste or repining."

To make the world a better place to live in is an ambition not falsified or diminished by the propensity of those who seek the reputation of having finely critical minds to say knowingly, "Ah, but what do you mean by better?" It is a philosophic naivety to suppose that a single, simple, uncontroversial and general form of words can represent all that is connoted by such particular declarations, not likely to be challenged individually, as that good drains are better than bad and that good books are better than bad, which are true even if people may not agree completely about what in a book entitles it to be classified in the one category or the other. Again, it is better to be well than sick and alive than dead. It is in a compendium of these and other possible particular meanings that we can say the purpose of science is to make the world a better place to live in and that —in Bacon's words—the "Dignity and proficiency" of science rest upon its ability to promote that wholly admirable ambition.

Notes

1. This is also the professional opinion: Charles Sanders Peirce, who was the principal American philosopher of mind, wrote: "the conclusions of science make no pretense to being more than probable," and John Venn said: "no ultimate objective certainty is attainable by any exercise of the human reason;" and Immanuel Kant: "Hypotheses always remain hypotheses, that is, suppositions to the complete certainty of which we can never attain."
2. *Logic, Semantics, Metamathematics: Papers from 1923 to 1928,* trans. J. H. Woodger (Oxford, 1956).
3. I well remember, when I was in Birmingham University, the derision with which a staff member in Eng. Lit. told me that the professor of physics in using the word "epitome" pronounced it in three syllables instead of four. Dear God, the fuss he made about it! He spoke in a voice as grave and hushed as he would have used if he had been telling me that the professor of rural economy, rightly supposing that his colleagues could not recognize a spade even if they saw one, had embezzled his department's funds and made off with them.
4. See Bryan Magee, *Towards Two Thousand* (London: McDonald, 1965).
5. In his novel *Headlong Hall* (London, 1816).

CAN SCIENTIFIC DISCOVERY

BE PREMEDITATED?

Introduction

Most scientists in most countries are funded directly or indirectly by the public purse. This puts scientists with social consciences under a special obligation to "improve natural knowledge" (in the Royal Society's words) or to procure the "advancement of learning" (in Bacon's). It is in this context that the question which forms the title of this essay is especially relevant. It is not rhetorical: I believe that it can be answered and that the answer has quite far-reaching political implications. I shall begin by recounting, with telegraphic brevity, three case histories of discoveries that could not possibly have been premeditated—that could not have been the specific outcome of a conscious and declared intention to make them. I shall then make some reference to the role of luck in scientific discovery, and shall finally turn to the question of whether our current understanding of how scientists make their discoveries is compatible with the idea of premeditation.

Case Histories

X-rays. [1] Thanks to anesthesia and to the development of aseptic surgery by W. S. Halsted of Johns Hopkins and Berkeley George Moynihan of Leeds, surgery had progressed so far and so fast by 1900 that Moynihan opined that not much further progress was to be expected.[2] Yet surgery still had one serious handicap: the surgeon had to embark on his operation without knowing what to expect inside. There was an imperative need for some method of making human flesh transparent. Imagine now that a system of research funding such as that which prevails today were in operation in 1900 and imagine also the incredulous derision with which a research proposal "to discover a means of making human flesh transparent" would be greeted by any grant-giving body. Yet as we all know, just such a procedure was discovered by a man primarily interested in studying electric discharges in high vacua. The medical potentialities of Roentgen rays (X-rays) for what is now called diagnostic radiology were recognized almost immediately.

HLA polymorphism. Second, imagine now a contract open to tender to discover the genetic constitutions predisposing human beings to any of three grave debilitating diseases: ankylosing spondylitis, multiple sclerosis and the juvenile (insulin-dependent) form of diabetes. No degree of premeditation could solve such a problem, but it was in fact solved, in the following way:

It took only a few years for the promising and exciting research on tumor transplantation started by C. O. Jensen's discovery around 1900 of tumor transplantability to reach a state of unparalleled confusion, with mutually contradictory and unreproducible results. Some of the early research workers of the Imperial Cancer Research Fund even began to speak of "seasonal" influences on the transplantability of tumors—a sure sign that research had reached a low ebb. The reason, we now know, is that the early research workers used different animals for each experiment—they used the "white mouse," the "brown mouse," the "spotted mouse" and even the "Berlin mouse" and the "Tokyo mouse," on the—at that time—virtually universal assumption that uniformity of color or provenance assured uniformity of reactivity. It was as if a chemist were to estimate the solubilities and melting points of "white chemicals" or "blue chemicals," etc., in the expectation that uniformity of color implied uniformity of physicochemical properties.

Research continued in a state of almost total disarray until matters were taken in hand by Peter Gorer of Guy's Hospital and Dr. C. C. Little and his colleagues, notably George Snell, J. J. Bittner and Leonell Strong, at Bar Harbor, Maine. Snell and Gorer between them worked out the genetic basis of tissue transplantation in mice and identified the chromosome segment housing the genes responsible for it: In mice, these form MHC, the so-called major tissue-matching complex, H-2.

The work of Gorer and Snell made it possible for Dausset and others to recognize and define the corresponding major tissue-matching system in man, known as HLA. It is sometimes thought that the major importance of the discovery of the HLA system was the facilitation of transplantation in man, but in my opinion its real importance was to have brought to light a new system of polymorphism in man—I mean a new system of stable genetic differentiation in the human population—one that has made it possible to specify the genetic constitutions predisposing their possessors to ankylosing spondylitis, multiple sclerosis and insulin-dependent diabetes, in much the same way that membership in the different blood groups is associated in different degrees to the susceptibility to gastric ulceration and gastric cancer. There is no conceivable means by which such a discovery could have been premeditated.

Nature of myasthenia gravis (MG). [3] Dr. Dennis Denny-Brown, a pupil and colleague of England's most famous neurophysiologist, C. S. Sherrington, is reputed to have been the first to draw attention to the similarity between the symptoms of the creeping neuromuscular paralysis myasthenia gravis and the symptoms of curare poisoning. This led directly to the use of anti-curare agents such as eserine (physostigmine) for the melioration of myasthenia gravis. The parallel surely implicates acetylcholine receptors of the postsynaptic muscle membrane— the area of apposition of muscle and nerve—in the causation of MG.

To investigate this possibility, Lindstrom and his colleagues (Lindstrom, 1979; Patrick and Lindstrom, 1973) decided to raise antibodies in rabbits against acetylcholine receptors.[4] Purified receptor protein from electric organs of eels was accordingly injected into rabbits, with the dramatic result that in due course they developed a flaccid paralysis whose similarity to myasthenia gravis was confirmed by the way they perked up after the injection of a muscle stimulant such as neostigmine.

These and subsequent findings (Newsom-Davis *et al.*, 1978) may be said to have corroborated the audacious hypothesis that myasthenia gravis is immunologically self-destructive in origin, as Simpson (1960) had brilliantly conjectured. There are other pieces to this jigsaw puzzle: MG is known to be accompanied by pathological changes in the thymus, the most important lymphoid organ in the body, which is often removed in the treatment of the disease. It is therefore very relevant that Wekerle and his colleagues (1978) have shown that embryonic precursors of muscle cells with well-formed acetylcholine receptors may differentiate from cells residing in the connective tissue framework of the thymus.

The special interest of these three examples is that although they represent discoveries which could not possibly have been premeditated—and could not therefore have been the subject of a customer/contractor treaty—they were nevertheless made by the ordinary processes of scientific inquiry, grossly inefficient and cost-ineffective though such processes have been declared to be by people who have no deep understanding of the nature of scientific research. All three examples do, however, illustrate the cardinal importance of the state of preparedness of mind, which is the subject of the next section.

Luck in Scientific Discovery

Any scientist who is not a hypocrite will admit the important part that luck plays in scientific discovery; this must always seem to be greater than it really is, because our estimate of its importance is inherently biased: We know when we benefit from luck, but from the nature of things, we cannot assess how often bad luck deprives us of the chance of making what might have been an important discovery (see p. 26)—the discoveries we did not make leave no trace. I think, therefore, that there was really no need for one of the world's most distinguished neurophysiologists to refer to his "feeling of guilt about suppressing the part which chance and good fortune played in what now seems to be a rather logical development" (Hodgkin, 1976).

It might nevertheless seem as if the principal lesson to be drawn from the three case histories I have just outlined is that luck plays a preponderant part in scientific discovery. I should like to challenge this view, for the following reasons having to do with the philosophy of luck:

We sometimes describe as "lucky" a man who wins a prize in a lottery at long odds; but if we describe such an event as luck, what word shall we use to describe the accidental discovery on a park bench of a lottery ticket that turns out to be the winning one?

The two cases are quite different. A man who buys a lottery ticket is putting himself in the way of winning a prize. He has, so to speak, purchased his candidature for such a turn of events and all the rest is a matter of mathematical probabilities. So it is with scientists. A scientist is a man who by his observations and experiments, by the literature he reads and even by the company he keeps, is putting himself in the way of winning a prize; he has made himself discovery prone. Such a man, by deliberate action, has enormously enlarged his awareness—his candidature for good fortune—and will now take into account evidence of a kind that a beginner or a casual observer would probably overlook or misinterpret. I honestly do not think that blind luck of the kind enjoyed by the man who finds a winning lottery ticket for which he has not paid plays an important part in science or that many important discoveries arise from the casual intersection of two world lines.

Nearly all successful scientists have emphasized the importance of preparedness of mind, and what I want to emphasize is that this preparedness of mind is worked for and paid for by a great deal of exertion and reflection. If these exertions lead to a discovery, then I think it would be pejorative to credit such a discovery to luck.

Methodology

Our present-day understanding of the methodology of science at bench or shop-floor level (as opposed to, for example, Thomas Kuhn's theory of the role of revolutions in the *history* of science) is something for which we are mainly indebted to Professor Sir Karl Popper, FRS. Popper's methodology is, I believe, quite incompatible with the idea that scientific discovery can be premeditated. Administrative high-ups in Washington and Whitehall firmly believe that scientists make their discoveries by the application of a procedure known to them as the scientific method—the belief in which, considered as a kind of calculus of discovery, is based on a misconception dating from the days of John Stuart Mill's *A System of Logic* and Karl Pearson's *The Grammar of Science.*

If such a method existed, none of us working scientists would be secure in our jobs, for consider a research worker in an institute devoted to elucidating the causes of and finding a cure for rheumatoid arthritis. If he fails to do so, his failure could only be either because he did not know the scientific method, in which case he should be sacked, or because he was too lazy or obstinate to apply it, an equally valid reason for dismissal.

There is indeed no such thing as "the" scientific method. A scientist uses a very great variety of exploratory stratagems, and although a scientist has a certain address to his problems—a certain way of going about things that is more likely to bring success than the gropings of an amateur—he uses no procedure of discovery that can be logically scripted. According to Popper's methodology, every recognition of a truth is preceded by an imaginative preconception of what the truth might be—by hypotheses such as William Whewell first called "happy guesses," until, as if recollecting that he was Master of Trinity, he wrote "felicitous strokes of inventive talent."

Most of the day-to-day business of science consists in making observations or experiments designed to find out whether this imagined world of our hypotheses corresponds to the real one. An act of imagination, a speculative adventure, thus underlies every improvement of natural knowledge.

It was not a scientist or a philosopher but a poet who first classified this act of mind and found the right word to describe it. The poet was Shelley and the word, *poiesis*, the root of the words "poetry" and "poesy," and standing for making, fabrication or the act of creation.

With this wider sense of the word in mind, Shelley roundly declared in his famous *Defence of Poetry* (1821) that "poetry comprehends all science," thus classifying scientific creativity with the form of creativity more usually associated with imaginative literature and the fine arts. What is more to the point is that Shelley went on to assert: "A man cannot say I *will* write poetry . . . the greatest poet even cannot say it."

No more, I submit, can a scientist say I *will* make a scientific discovery; the greatest scientist even cannot say it.

Summary and Conclusions

I began this essay with three case histories of discoveries that could not possibly have been premeditated and that could therefore never have been the subject of customer/contractor treaties. I then went on to say that it would be injudicious to credit to luck the consequences of a conscious preparedness of mind. Finally, I argued that the modern conception of scientific procedure taken in conjunction with Shelley's conception of poetry is incompatible with the idea that scientific discovery could be premeditated. In short, I believe I should now be in a position to answer the question posed in the title: "Can scientific discovery be premeditated?" The answer is: "No."

Notes

1. This example, of the medical uses of X-radiography, was put into my head by my old teacher Dr. John Baker, FRS (1942), via a lecture by Sir John McMichael, FRS.
2. Under the pseudonym "A Harley Street surgeon," Moynihan first expressed this view in *The Strand* magazine about 1900. He expressed it again in a publication of Leeds University Medical School in 1930 and yet again in his Romanes Lecture in Oxford University for 1932. See P. B. Medawar: *Pluto's Republic* (Oxford, 1982); pp. 298–310.
3. For this example, I am indebted to a clinical lecture delivered in the Clinical Research Centre (Harrow) by Dr. John Newsom-Davis (January 1980).
4. Acetylcholine is the chemical that causes a motor nerve impulse to initiate muscular contraction.

References

Baker, J. R. 1942. *The Scientific Life.* London: Allen & Unwin.

Hodgkin, A. L. 1976. Chance and design in electrophysiology: an informal account of certain experiments on nerve carried out between 1934 and 1952. *J. Physiol.* 263: 1–21.

Kuhn, T. S. 1979. *The Structure of Scientific Revolutions.* Chicago.

_____. 1978. *Essential Tension.* Chicago.

Lindstrom, J. 1979. Autoimmune response to acetylcholine receptors in myasthenia gravis and its animal model.

Newsom-Davis, J., A. J. Pinching, Angela Vincent, and S. G. Wilson. 1978. Function of circulating antibody to acetylcholine receptor in myasthenia gravis: investigation by plasma exchange. *Neurology* 28: 266–72.

Patrick, J., and J. Lindstrom. 1973. Autoimmune response to acetylcholine receptor. *Science* 180: 871–72.

Pearson, Karl. 1892. *The Grammar of Science.* London.

Popper, K. R. 1959. *The Logic of Scientific Discovery.* London.

Simpson, J. A. 1960. Myasthenia gravis: a new hypothesis. *Scottish Medical Journal* 5: 419–35.

Whewell, William. 1840. *The Philosophy of the Inductive Sciences.*

THE LIMITS

OF

SCIENCE

It is important to realize that science does not make assertions about ultimate questions—about the riddles of existence, or about man's task in this world. This has often been well understood. But some great scientists, and many lesser ones, have misunderstood the situation. The fact that science cannot make any pronouncement about ethical principles has been misinterpreted as indicating that there are no such principles while in fact the search for truth presupposes ethics.

—KARL POPPER
Dialectica 32: 342

O Timothy, keep that which is committed to thy trust, avoiding profane and vain babblings, and oppositions of science falsely so called: which some professing have erred concerning the faith.

—1 TIMOTHY 6: 20–21

Abstract

The "new philosophers" of the seventeenth century in England did not clearly envisage a limit to science. Their motto (the Spanish origin of which will be described here) was *Plus Ultra*. In science, they believed, there would always be more beyond. The existence of a limit to science is, however, made clear by its inability to answer childlike elementary questions having to do with first and last things—questions such as "How did everything begin?" "What are we all here for?" "What is the point of living?" Doctrinaire positivism dismisses all such questions as nonquestions or pseudoquestions—hardly an adequate rebuttal because the questions make sense to those who ask them, and the answers to those who try to give them. This essay tries to explain why science cannot answer these ultimate questions and why no conceivable advance of science could empower it to do so. The author considers but finds fault with the idea that the growth of scientific understanding is self-limited, that is, is slowed down and brought to a standstill as a consequence of the act of growing itself, as with the growth of populations, skyscrapers or aircraft.

There is an intrinsic, built-in limitation upon the growth of scientific understanding. It is not due to any cognitive incapacity. It is a logical limitation that turns on a "Law of Conservation of Information."

It is not to science, therefore, but to metaphysics, imaginative literature or religion that we must turn for answers to questions having to do with first and last things. Because these answers neither arise out of nor require validation by empirical evidence, it is not useful or even meaningful to ask whether they are true or false. The question is whether or not they bring peace of mind in the anxiety of incomprehension and dispel the fear of the unknown. The failure of science to answer questions about first and last things does not in any way entail the acceptability of answers of other kinds; nor can it be taken for granted that because these questions can be put they can be answered. So far as our understanding goes, they can not.

Reasons are given, though, for supposing that there is no limit upon the power of science to answer questions of the kind science *can* answer. This is science's greatest glory, for it entails that everything which is possible in principle can be done if the intention to do it is sufficiently resolute and long sustained.

1   *Plus Ultra?*

Plate 1 is a reproduction of the frontispiece of *Novum Organum* (1620), the first work of Francis Bacon's projected System of Philosophy, *The Great Instauration*. It portrays the Straits of Gibraltar flanked by the colossal pillars of Hercules, each so enormous as to reduce the Empire State Building to the scale of an outdoor privy. Beneath the pillars is an inscription (Daniel 12: 4), regarded as deeply prophetic and portentous by the "new philosophers" ("scientists," we now say) of Bacon's generation, many of whom were in holy orders. The inscription runs: "Many shall pass to and fro and knowledge shall be increased"—a declaration deemed to presage the great voyages of discovery of the period, the movement of people to and from Britain and the continent of Europe, and especially from Britain to North America. It was deemed also to foretell the advancement of learning of which Bacon was the principal evangel.

These pillars played a very important symbolic part in the great scientific revolution in the age of Francis Bacon and Jan Amos Comenius. I have quoted elsewhere[1] Professor Marjorie Hope Nicolson's account of the matter as she gave it in a lecture delivered at Rockefeller University:

Plate 1: Frontispiece of Francis Bacon's *The Great Instauration*.

Before Columbus set sail across the Atlantic, the coat of arms of the Royal Family of Spain had been an *impresa*, depicting the Pillars of Hercules, the Straits of Gibraltar, with the motto, *Ne Plus Ultra*. There was "no more beyond." It was the pride and glory of Spain that it was the outpost of the world. When Columbus made his discovery, Spanish Royalty thriftily did the only thing necessary: erased the negative, leaving the Pillars of Hercules now bearing the motto, *Plus Ultra*. There was more beyond.

This makes a good story, though such frugality was not to be expected of a ruling family born of the union of the houses of Aragon and Castile; but Professor Earl Rosenthal's reexamination of the matter[2] makes it clear that the account needs emendation. Although the motto was traditionally associated with them, the pillars never actually bore the inscription *Ne Plus Ultra*. That Hercules himself inscribed these words upon the columns is as legendary as the columns themselves, but that the columns were the heraldic emblems of the Spanish monarchy is indubitable. It is true also that upon the discovery of America, royalty erased the negative so that the motto became *Plus Ultra:* It was now the pride and glory of Spain that so far from being the outpost of the old world, Spain was the gateway opening upon a new world of who could know what riches and promise of adventure. Plate 2, showing a choir stall in Barcelona Cathedral, illustrates the heraldic emblem of King Charles I of Spain (the Holy Roman Emperor Charles V), the cloaks of whose German guards came to be known as "plus ultras" because they were embroidered with the regal emblem.

Plus Ultra was seized upon as a motto or slogan by the pioneers of science at the end of the sixteenth and beginning of the seventeenth centuries, and it was a slogan admirably well suited to what they hoped of the new philosophy. There was no

Plate 2: A choir stall in Barcelona Cathedral. (From a photograph taken by the agency Arxiu Mas, Barcelona.)

limit to science; there would always be more beyond. Nor were they wholly mistaken. In terms of the fulfillment of declared intentions, science is incomparably the most successful enterprise human beings have ever engaged upon.[3] Visit and land on the moon? A fait accompli. Abolish smallpox? A pleasure. Extend our human life span by at least a quarter? Yes, assuredly, but that will take a little bit longer.

The Reverend Dr. Joseph Glanvill (1636–1680) wrote a paean for the new science titled *Plus Ultra* (1668), in which in addition to the fashionable swipes at Aristotle he advised the new philosophers upon how to write and in what spirit to conduct their experiments.

Henry Power, FRS, was so far confident of the new science as to say (in *Experimental Philosophy,* 1664): "The progress of art is indefinite and who can set a non-ultra to her endeavours?" (By "art," of course, Power meant what we should today call engineering or craft.)

The great question that forms the subject of this essay is: Were these sanguine and eager pioneers right; is it always to be Plus Ultra; is there no limit upon the advancement of scientific understanding and the power it confers upon us?

Because he was the great spokesman of the new philosophy, Bacon should be allowed to speak first. There are passages in Bacon (I think especially of the Preface to *The Great Instauration*) which give the impression that Bacon believed that only a failure of nerve could retard the progress of science. Thinking of the Pillars of Hercules which were his frontispiece and the *Ne Plus Ultra* that had traditionally gone with them, he wrote (Gilbert Wats's translation): ". . . sciences also have, as it were, their fatal columns; being men are not excited, either out of desire or hope to penetrate farther."

However, in Book I of *De Dignitate et Augmentis Scien-*

tiarum, Bacon, writing explicitly upon science, and conceding that science might "comprehend all the universal nature of things," spoke of three limitations:

> The first, that we not so place our felicity in knowledge as we forget our mortality: the second, that we make application of knowledge to give ourselves repose and contentment and not distaste or repining: the third, that we do not presume by the contemplations of nature to attain the mysteries of God. [Trans. Wats, 1674.]

The Ultimate Questions

That there is indeed a limit upon science is made very likely by the existence of questions that science cannot answer and that no conceivable advance of science would empower it to answer. These are the questions that children ask—the "ultimate questions" of Karl Popper. I have in mind such questions as:

> How did everything begin?
> What are we all here for?
> What is the point of living?

Doctrinaire positivism[4]—now something of a period piece—dismissed all such questions as nonquestions or pseudoquestions such as only simpletons ask and only charlatans of one kind or another profess to be able to answer. This peremptory dismissal leaves one empty and dissatisfied because the questions make sense to those who ask them, and the answers, to those who try to give them; but whatever else may be in dispute, it would be universally agreed that it is not to science that we should look for answers. There is then a prima facie case for the existence of a limit to scientific understanding.

Of What Kind Might This Limit Be?

I shall consider two major possibilities, each of which can be further subdivided:

1. The growth of science is self-limiting—that is to say, is slowed down and eventually brought to a standstill as a consequence of the process of growth itself—just as the growth of, for example, natural populations, skyscrapers and aircraft is inevitably restrained by the consequences of the process of growing.

2. As an alternative possibility, there might be some *intrinsic* limitation upon the growth of scientific understanding, and this in turn might be of two kinds.

2.1 *Cognitive*—that is, having to do with apprehension and the input side of awareness, as I shall try to illustrate by a parable having to do with the intrinsic limitations of the resolving power of an ordinary light microscope.

2.2 *Logical*—that is, arising out of the very nature of ratiocination. I shall suggest that to expect science to answer the ultimate questions is tantamount to expecting to deduce from the axioms and postulates of Euclid a theorem having to do with how to bake a cake.

2   Is the Growth of Science Self-Limited?

In the last chapter, I referred to the self-limitation of growth as that which might arise as a direct consequence of the process of growing and gave the growth of populations and of skyscrapers as examples. Consider these first. Populations are in principle capable of growing by continuous compound interest because, as with biological growth generally, that which results from growth is itself capable of growing. In real life, no population can grow exponentially for more than a few generations: The growth rate of all real populations is limited by the action of one or more density-dependent factors. Among such factors are the depletion of food, the accumulation of waste and the effects upon reproduction of the psychophysiological wounds of the stress brought about by overcrowding—all of which conspire to keep the size of a population well below its Malthusian potential.

It is nonsense, therefore, to suppose that a human population could ever become so numerous as to be standing shoulder to shoulder upon the land surface of the earth. It is by no means nonsense, though, that unless the birth rate drops to a level commensurable with the death rate, the density-dependent factors that arrest the growth of the population will include the famine and pestilence expected of the Malthusian apocalypse.

This is no mere theoretic threat, such as the threatened heat death of the universe when entropy has done its work. No; in Mexico City and the Horn of Africa today we can already hear the opening bars of the *Dies Irae* in that requiem for mankind which may be written to lament the consequences of legislative negligence and the systematic depreciation, for religious or political reasons, of the magnitude of the threat that now confronts us.

I mentioned the growth of skyscrapers as a second example of self-limitation of growth. Considering how much civic pride in America is invested in having a skyscraper taller than anyone else's, people may wonder why a skyscraper may not grow as high as a city's citizens may wish. The answer is obvious when pointed out: unless the upper stories are to remain uninhabited, the proportion of floor space allocated to elevators in the lower stories soon becomes grossly uneconomic—a truth of which we are vividly reminded by those tall buildings on which the elevators crawl up outside.

A third example of self-limitation of growth turns upon the geometric truism often referred to (wrongly, I have been told) as Spencer's Law, which affirms that if a three-dimensional body grows in size without change of shape, its surface area increases as the square of a linear dimension while its volume or mass increases as the cube. This principle sets an upper limit[1] upon the sizes of land animals. The weight of an animal increases as the cube of a linear dimension, while the power of the legs to support the weight depends on the cross-sectional area of the limbs, which increases only as the square. An elephant's legs are pillar-like already. If an elephant were much larger than it is, it might be difficult to see daylight between its legs. The same principle applies to jumbo aircraft. Whatever the commercial and aerodynamic advantages of growing still larger, jumbo jets have a problem with the degree of massiveness of landing gear that must support the aircraft's weight on the impact of landing. So far, happily, metallurgy has advanced fast enough to keep pace with these demands upon the landing gear.

With these four examples of self-limitation of growth in mind, we may now ask what forms such a constraint might take in the context of the advancement of learning. The following possibilities have often been canvassed:

1.   Scientific knowledge is now so hugely voluminous that a scientist can no longer get his bearings in a voyage of discovery: No one can know what is already known and what still remains to be found out.

2.   Science is now so specialized and so fragmented that no synthesis is possible, and therefore no advance, for which such a synthesis would be required.

3.   Science at its frontiers is so far advanced and its concepts so difficult to grasp that it is now beyond the comprehension of any one mind. A scientist cannot henceforward expect to qualify himself for research by a conventional four plus three years of tertiary education: ten or twelve years of study will be needed from now on to equip a scientist to take his place in the front line.

4.   A fourth possibility is the most alarming and for that reason the most widely canvassed: Science has now so far outgrown our moral stature that we shall all destroy ourselves in a Promethean fire[2] of our own making or be damned in a huge Faustian payoff for the sheer impiety of seeking to find out that which might with advantage have remained unknown.

Is there not a measure of truth in each of these?

In my opinion there is not:

1. The problems raised by the—as it sometimes seems—oceanic volume of scientific knowledge are essentially technological problems, for which adequate technological solutions are rapidly being found; computers like those used in the great medicobiological extracting services such as *Excerpta Medica* of Amsterdam endow a medical scientist with an almost infinitely capacious exosomatic[3] memory with prompt and trustworthy powers of information retrieval.

2. Science is now so specialized and fragmented that no synthesis is possible and therefore any advance for which such a synthesis would be required cannot now take place. So tiny and so far apart are the smallholdings we cultivate that scientists nowadays can hardly communicate with each other—but then, there never was *one* science only, no single shapely whole that we lesser mortals cut up smaller and smaller to make it easier to cultivate. There were always sciences and there were always arts, and no one man knew them all—no one man ever had the know-how to make glass, brew beer, dress leather, make paper and cast a bell.

And if it were indeed true that one biologist can hardly communicate with another or one physicist with another, must it not be true a fortiori that so deep a chasm separates physicists on the one side from biologists on another that they might work on different planets, for all the intercourse there could be between them?

This ought to be true, it is felt, but what is in fact true is something altogether different: So far from contemplating the biological sciences in a bemused stupor of incomprehension, chemists and physicists have entered the world of biology and given us a new and deeper understanding of the structures and performances of living creatures—especially their growth and heredity. If specialization had already gone to the extremity that the duller and more boring speechmakers allege, and if biologists and physicists could communicate only in dumbshow, there would be no molecular biology today. Sciences are becoming more unified, not less: Science approximates more and more closely to that shapely whole which was supposedly its beginning.

The reason why actively creative scientists seem so seldom to communicate with each other is that they do not really want to. Scientists whose work is prospering are wrapped up in it with such obsessive absorption that they want above all to be left to cultivate their gardens. They are not more than idly curious about what goes on in other people's gardens, and their closest approximation to neighborly behavior often amounts to little more than an inclination to borrow their neighbors' garden tools—especially the physicists'.

3. The next conceivable process of self-limitation to be considered is that which turns on the familiar complaint that at its most forward frontiers, science is at the very limit of comprehensibility. A scientist henceforward must train for ten to fifteen years if he is to take his place in the front line of those engaged in the struggle for understanding, and even as it is, modern science is beyond the comprehension of any one mind.

This might be a very telling complaint if it were single minds that we ever relied upon, but in reality we work not by single minds but by consortia of intelligences, past as well as present; for what we think or do now is a function of what others have thought and done before us—people whose past findings and past errors are part of our own inheritance of understanding. Thus a television set (perhaps the most complicated science-based contraption in everyday use) is not within the effective comprehension of any one mind, for there is no one person who knows the electronics and the glass and vacuum technology and has the know-how of plastic molding to such a degree of proficiency that if some holocaust were to obliterate science and technology so that we had to begin again, this one knowledgeable human being could reinstruct and redirect the activities of those who would in due time reconstruct a television set. Clearly it was a committee or consortium of engineers and technologists, not any one man, that had the theoretic understanding and practical know-how to build a TV set. It was a great cooperative enterprise that brought TV sets into being, unlikely though it may at one time have seemed.

As to the scientist's being now obliged to spend ten to fifteen years before he can become adequately proficient in research, the scientist takes much longer than that already, for what is research but learning—and what scientist ever feels that, being complete, his research is now at last finished? The nature of science is such that a scientist goes on learning all his life—and must—and exults in the obligation upon him to do so. There is no determinate process of education at the conclusion of which a scientist can flex his muscles and pronounce himself ready at last to take part in the long struggle against ignorance and disease. It would be different if science had a goal that could be attained, but it has not. As explained and justified on pp. 5, 41 n. 1, there can in science be no apodictic certainty—that is, no finally conclusive certainty beyond the reach of criticism.

4. I turn finally to the last of the four agencies that might, I suggested, retard and eventually halt the advancement of learning. I have in mind the great Promethean fire which, it is feared, will one day do us all in.

This is not a valid example of self-limitation of growth, because such a denouement is in no way logically entailed by anything that has gone before. It is of course theoretically conceivable that an inmate of some institution of genetic engineering, fired by a lunatic variant of the Baconian ambition embodied in "The Effecting of All Things Possible"—the ambition to do something simply because it *can* be done—might reclothe the nucleic acid of highly virulent virus with a non-antigenic protein, so depriving the body of any defense against it. Now, that really would be the end of us. It is conceivable, I say, but it is not logically entailed by anything that has happened before in the growth of virology or molecular genetics, the practitioners of which are second to none in sobriety and solemnity of purpose. In real life, no one behaves like the wicked scientists of gothic science fiction movies any more than anybody ever did behave like dramatis personae of the writings of Mary Shelley or Mrs. Ann Radcliffe. Should it not be part of the common law of the world of learning that its citizens could be judged sane until shown to be otherwise?

3   Is There an Intrinsic Limitation upon the Growth of Science?

B efore embarking on these discussions, I put it to the reader that the limitations upon science might be of two kinds, each capable of further subdivision. The first possibility was the self-limitation of growth—that the growth of science is of logical necessity retarded by the consequences of the act of growing —but having considered four possibilities, I concluded that there was no such limitation upon science.

The second possibility was the existence of an intrinsic limitation upon science, whether through either a cognitive inadequacy or a restriction arising out of the very nature of the process of reasoning. "Cognitive inadequacy" needs to be explained.

The Parable of the Microscopist

The "strength" of a microscope is not a function of its power to magnify the object studied, for there is no practical limit to magnification and it might well be "empty"—i.e., not such as to reveal further detail.[1] It is *resolving power* the microscopist is after—that is, the power to tell apart objects that lie very close together. By the middle of the nineteenth century, the compound light microscope had become an instrument fine enough to arouse ambitions it could not itself fulfill. The microscopist was made aware of the existence of a world of fine structure which it was not within his power to make out: the world of objects of the nanometer[2] and order of size. This was the world of viruses, of the smallest microorganisms, of cellular and nuclear organs such as mitochondria and chromosomes, the world also of such protozoa as foraminifera and diatoms, the chalk organisms, often with shells of the most exquisitely fine engravery and a rococo profusion of fine ornamentation. Though known to exist, most of this world lay just outside the microscopists' reach. (Plate 3 illustrates some of this fine detail.)

Under the pressure of demand from students of cell structure, bacteriologists and many gifted amateurs—men such as Brewster, Wollaston, Coddington and above all Ernst Abbé of Jena—improved the microscope mightily, adding to it a substage condenser, which focused light upon the object to be examined, and achromatic and apochromatic lenses, the latter correcting for both chromatic and spherical aberration. This all helped, but still it wasn't good enough: The farther forward the frontier of the visible was pushed, the more tantalizing and inaccessible became the world that still lay out of sight. This was the ultramicroscopic world, and its inaccessibility was due to an intrinsic limitation upon the resolving power of the light microscope: It is not possible to resolve, to tell apart, two structures closer together than about half the wavelength of visible light. No amount of peering down a light microscope and no effort of attention could now avail; there was an intrinsic cognitive barrier.

An analogous intrinsic limitation is imposed on a fisherman who tries to catch fish smaller than the apertures of his net; except by luck, he is quite unable to do so.

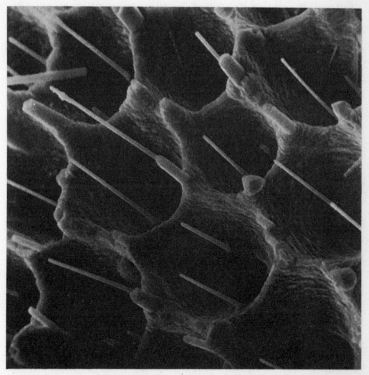

Plate 3: A scanning electron photomicrograph of the shell of a foramini-
feron *Globigerinoides sacculifera*, showing detail that cannot be
resolved by ordinary light microscopy. (Taken by Mr. H.A. Buckley
of the British Museum [Natural History] and reproduced here
by courtesy of the Trustees.)

The question is, are we scientists in a situation analogous to the microscopist's—are our senses (using that word in the widest sense) intrinsically unable to enjoy access to a source of information needed to answer the ultimate questions? The implied analogy is not a sound one. The microscopist knew he was limited, knew why he was so, and could take steps to remedy his condition. We do not and cannot. Among the microscopists' remedies were the substitution for visible light of light of shorter wavelength, such as ultraviolet light used in conjunction with quartz lenses, which are permeable by it, and using photography in place of direct vision. The end point was of course the substitution of an electron beam (Plate 3) for light and of a fluorescent screen for direct inspection: "electron microscopy," which uses magnets instead of lenses to focus the beam. Thus, having recognized the nature of the limitation upon him, the microscopist could find remedies for the constraint upon him, as scientists have not and cannot.

The Law of Conservation of Information

Lovers of Lewis Carroll's two *Alice* books, long recognized as a mathematico-philosophic satire, will remember the poignant scene in which the Walrus and the Carpenter invite some oysters for a walk along the beach, with a view to making a meal of them. To distract their attention from this ambition, the Walrus addressed them as Carroll recounts:

> "The time has come," the Walrus said,
>   "To talk of many things:
> Of shoes—and ships—and sealing-wax—
>   Of cabbages—and kings—
> And why the sea is boiling hot—
> And whether pigs have wings."

The Walrus might have saved his breath by telling the oysters that the time had now come to discuss a number of topics of empirical observation and certain problems arising out of them.

The relevance of this unexpected divagation into poetry will become clear in expounding the Law of Conservation of Information, which runs as follows: *No process of logical reasoning —no mere act of mind* or *computer-programmable operation— can enlarge the information content of the axioms and premises or observation statements from which it proceeds.*

In lay usage, "information" is thought of as an abstract concept concretely exemplified by some such proposition as "Madrid is the capital of Spain," but in a professional vocabulary, "information" connotes structure or orderliness, especially of the kind that makes possible the transmission of a meaningful message or in the form of a communication that prescribes and confers specificity upon any structure or performance. Thus the information structurally encoded in the huge molecule of deoxyribonucleic acid (DNA) is such as to specify the development of this particular organism and not that, and the wealth of information embodied in an architect's blueprint specifies some one building and no other. I attempt no demonstration of the validity of this law other than to challenge anyone to find an exception to it—to find a logical operation that will add to the information content of any utterance whatsoever.

We can illustrate the working of the law by the theorems of Euclid's geometry. Although sometimes unfamiliar and unexpected (the theorem of Pythagoras surprised Thomas Hobbes into the utterance of an oath), the theorems of Euclid are merely a spelling out, a bringing into the open, of information already contained in the axioms and postulates. Given the axioms and postulates, to a perfect mind (as A. J. Ayer remarked), the theorems of Euclid would be instantly obvious, without the necessity for making the information they contained explicit by a complicated deductive derivation. Indeed, philosophers and logicians since the days of Bacon have been entirely clear on this point: deduction merely makes explicit information that is already there. It is not a procedure by which new information can be brought into being.

I must now try seriously to find fault with the Law of Conservation propounded above. Consider first a so-called inductive law, such as the philosophers' favorite, "All swans are white," or one of Max Beerbohm's famous inductive laws, such as "Young men with prematurely gray hair are invariably charlatans," or alternatively the law formulated by P. G. Wodehouse's famous alienist, Sir Roderick Glossop: "A lay interest in matters to do with liturgical procedure is invariably a prelude to insanity." The potential challenge to the Law of Conservation lies in the question of the process of reasoning that empowers us to pass from empirical awareness that this, that and the other swans are white to affirm the truth of the law that *all* swans are white. And the same question may of course be asked of how we arrive at Beerbohm's law from an awareness that this, that and the other gray-headed young men are charlatans. The answer is that of course there is *no* process of logical reasoning by which we can proceed from these respective sets of particulars to the general laws that embody them, and no philosopher ever claims that such generalizations are more than guesses; indeed, if there *were* such a process of reasoning, it would certainly flout the idea of conservation of information. No inductive generalization can contain more information than the sum of its known instances. An inductive law is a hypothesis that has no claim whatsoever to certainty.

In no sense, then, do inductive "laws" constitute a threat to the conservation law I am trying to find fault with; nor is the existence of the computer program such as Dr. Pat Langley's Bacon 3 as he recounted it at the sixth international joint conference of artificial intelligence. Upon being fed the right empirical data, Bacon 3 rediscovered Boyle's gas law and Kepler's third law of planetary motion by searching in the data for correlations and invariances. This, of course, as Langley makes clear, is not discovery but rediscovery: The information fed into the computer obeyed the laws that came out of it. What the computer program achieved was exactly parallel to that which deduction achieves: namely, to bring out and make explicit the information already contained in the data. It would be regarded as a great comic episode if upon being fed the kind of information in which were discovered Boyle's law and Kepler's third law the computer were to come up with a formulation of Le Chatelier's theorem.

Now at last we may ask what is the relevance of all the rigmarole about oysters to the problem of finding answers to the ultimate questions. The propositions and observation statements of science have empirical furniture only: In epistemological principle, they have all to do with ships, shoes and sealing wax, etc. To say as much is not in any way to diminish science, for the material world is full of wonderful and inspiring things. Some are commonplace and ordinary, to be sure—among them perhaps raindrops, pebbles and water fleas—but others are awe-inspiring or in the literal sense tremendous: the multitudinous seas round Cape Horn, where the Atlantic and Pacific oceans fight it out to see which is the greatest; and the great dome of heaven, as we may become aware of it on a high plateau, for nowhere else does the world seem so large. All are part of the empirical mise-en-scène of the world.

The Law of Conservation of Information makes it clear that

from observation statements or descriptive laws having only empirical furniture there is no process of reasoning by which we may derive theorems having to do with first and last things; it is no more easily possible to derive such theorems from the hypotheses and observation statements with which science begins than it is possible to deduce from the axioms and postulates of Euclid a theorem to do with how to cook an omelet or bake a cake—accomplishments that would at once unseat the Law of Conservation of Information. I do not believe that revelation is a source of information, though I acknowledge that it is widely believed to be so—and that Coleridge judged theology Queen of the Pure Sciences for that very reason.

4   Where *Plus Ultra* Prevails

It is implicit in the foregoing arguments that a distinction must be drawn between questions of the kind science can answer and questions belonging to some other world of discourse, to which we must turn instead if they are to be answered at all.[1] I suppose, though, that the question a person of brusque and pragmatic temper would be most likely to put and most likely to want answered is this: Is there any limit to the power of science to answer questions of the kind that science *can* answer? I do believe that for methodological reasons, the answer is surely not: In the world of science, the "fatal columns" Bacon spoke of (p. 65) do not define a limit we cannot go beyond. In science as such there is always more beyond. I think this inference can be drawn from a brief consideration of the nature of the creative act in the advancement of science.

Although some diehard inductivists still believe, as John Stuart Mill did, that there may be propounded a calculus of discovery—a formulary of thought which can conduct us from observation statements to general truths—most methodologists, however much they may differ in other ways, believe that the generative act in science is the brainwave, inspiration or flash of imaginative insight that is the propounding of a hypothesis, a hypothesis being always an imaginative preconception of what the truth might be. William Whewell, we have seen, at first described hypotheses as "happy guesses," though later the then holder of the most prestigious academic post in England spoke—ahem—of "felicitous strokes of inventive talent." Hypotheses are of course imaginative in origin. It was not a scientist or a philosopher but a poet who first classified this act of mind and found the word for it. As I explained earlier, in a more general context (p. 52), the imaginative exploit that generates a scientific hypothesis was regarded by Shelley as cognate with poetic invention. He was using the word "poetry" in the root sense *poiesis*—the act of making, of creation. Certainly hypotheses are products of imaginative thinking.

Ever since Plato spoke of the divine rapture or divine fury of creativity, the act of poetic invention has been held in awe by those who exercise it, just because it seems to embody an infringement of divine copyright—a making something of sense of beauty or creating order out of nothingness. Samuel Taylor Coleridge wrote in his *Biographia Literaria* (1817): "The primary imagination I hold as . . . a repetition in the finite mind . . . of the eternal act of creation[2] in the infinite I AM." Descartes regarded imagination as a faculty of the soul, therefore as "a wind, a flame or an ether"; and in some renderings of the myth of Prometheus, the stolen fire made possible the faculty of creativity as it expressed itself in the arts and sciences. The world's great creative geniuses have been thought of as so many flames of this Promethean fire. Writing on imagination, Dr. Samuel Johnson was far from pentecostal. The delight of intercourse between the sexes, he opined, was due chiefly to imagination: "Were it not for imagination, sir, a man would be as happy in the arms of a chambermaid as of a Duchess" (*Boswell's Life of Johnson*, eds. G. B. Hill and L. F. Powell, vol. 3 [Oxford, 1934], p. 342). We should have liked to have heard more of

Johnson's imagination, but "it would not be proper," said Boswell, "to record the particulars of such a conversation in moments of unreserved frankness." For Wordsworth, "imagination is but another name for "clearest insight, amplitude of mind and reason in her most exalted mood."

If the generative act in science is imaginative in character, only a failure of the imagination—a total inability to conceive what the solution of a problem *might* be—could bring scientific inquiry to a standstill. No such failure of the imagination—nor any failure of nerve that might be responsible for it—has yet occurred in science and there is not the slightest reason to suppose that it will ever do so. A drying up of science is no more easily conceivable than a drying up of musical creation or imaginative literature. People who are fed up with modern music or modern literature are essentially people who think that creativity should have taken a different turn; there is no thought in their minds that creativity as such has come to a standstill. They complain only of that which is created and do not question the continued existence of a creative faculty.

Methodologists differ in their interpretation of the process of evaluating hypotheses. Although evaluation turns on matching hypothesis with real life (see "Truth," pp. 4–6), Thomas Kuhn[3] regards the testing of hypotheses not as a private transaction between scientist and reality but rather as a matter of measuring the hypothesis against the prevailing "paradigm"— the prevailing orthodoxy or establishment of received beliefs and ways of thinking. Certainly most of the day-to-day business of science consists in making observations and doing experiments bearing upon the acceptance or modification of hypotheses.

Because an analogous question must eventually be asked of candidate answers to the "ultimate questions," we have now to ask what characteristic of a hypothesis justifies its being considered at all to be scientific—i.e., to belong to the domain of science and common sense. Kant's answer was that it must be unconditionally true of any hypothesis that it could *possibly* be true—which I think was Kant's way of formulating the demarcation criterion referred to by thinkers since as "verifiability in principle" or "falsifiability in principle," correspondence or noncorrespondence with reality being the essence of them all.

My contention in the discussion so far has been that it is logically outside the competence of science to answer questions to do with first and last things. Addressing now a different question, I urge that there is no limit upon the ability of science to answer the kind of questions that science *can* answer. Never once in the history of science have we reached a *Non Ultra;* nothing can impede or halt the advancement of scientific learning except a moral ailment such as the failure of nerve Bacon (p. 65) had in mind when he wrote of those who would not pass beyond the fatal columns into the open seas that lay beyond them—and is it seriously conceivable that some philosophic ailment could dry up the imagination that begets new scientific ideas? If so, we should have had a premonition of it in the drying up of more ancient imaginative endeavors such as literature, music making and the fine arts generally—and the only people who might think such an event at all likely are those whose own inventiveness has for some reason, probably of no general significance, dried up (something that happens to us all sometimes), but surely none of us can have been so arrogant or lacking in humor as to suppose this a world malady. Science will persevere just as long as we retain a faculty we show no signs of losing: the ability to conceive—in no matter how imperfect or rudimentary a form—what the truth *might* be and retain also the inclination to ascertain whether our imaginings correspond to real life or not.

Science could of course be brought to an end by a catastrophe. Science cannot be prosecuted in a radiation-blasted wasteland—but if science were to come to an end for *that* reason, it would be one episode only, and not the most important, in a much more awful tragedy. Catastrophe apart, I believe it to be science's greatest glory that there is no limit upon the power of science to answer questions of the kind science *can* answer.

That which is science's greatest glory is also, unhappily, its greatest threat: for taking "possible in principle" to signify "not being such as to flout the second law of thermodynamics" or any other bedrock physical principle, what is being said is, in effect, that in the world of science anything that is possible in principle can be done if the intention to do it is sufficiently resolute and long sustained. This places upon scientists a moral obligation which, considered as a profession, they are only just now beginning to grapple with. From our political masters it calls for a degree of wisdom, scientific understanding, political effectiveness, world sense and goodwill that no administration in any country has yet been able to muster.

5 Where Then Shall We Turn?

I ended the last chapter with the brave claim that there was no limit upon the power of science to answer questions of the kind that science can answer. If the ultimate questions can be answered—something I have no *reason* to believe—we must seek transcendent[1] answers, by which I mean answers that do not grow out of or need to be validated by empirical experience: answers that belong to the domains of myth, metaphysics, imaginative literature or religion. But, I was once asked, "might not the answer to a question such as 'How did everything begin?' be empirical in character?"

I believe not. We can hardly have empirical awareness of a frontier between being and nothingness without also having an empirical awareness of what lies on either side of it, and whereas the hither side of the frontier poses no special problem, for we can be empirically aware of that which is in being, there can be no empirical awareness of nothingness, so that if any such frontier exists it cannot exist in the domain of discourse of science and common sense.

(Voice from the back of the hall: "Give an example of what you describe as 'imaginative literature' as a fount of understanding to help answer your questions.") What I had in mind was the remarkable declaration known to us as Genesis 1:2: ". . . and the

Spirit of God moved upon the face of the waters." I remember from my school days how very many of us were surprised by the occurrence in the Bible of a statement so distant from the idiom of the Pentateuch, for the books of the Pentateuch are generally speaking rather unfanciful in character, being rather grittily matter-of-factual and particular in their narrative of divine propositions and dispositions. I described the phrase from Genesis as imaginative literature because it is so far removed from the businesslike narrative of the remainder. Like very many others, I have been awed and rather moved by the notion it embodies and may have felt, too, a certain accession of understanding of a visceral rather than an intellectual kind. If Shakespeare or Bacon had used the image, we should have thought it wonderfully moving and meaningful—an example, perhaps, of poetic truth as opposed to that drab literal truth by which scientists are preoccupied.

Judging from the reactions of audiences to whom I have lectured upon the limits of science, people are not as a rule mystified by my reference to seeking in myth, metaphysics or religion for answers to questions about first and last things.

As for myth, I am not at all in sympathy with those modern anthropologists who regard myth and science as alternative explanatory stratagems of the same stature, though independent and of different origins. Myths have often a rich quasi-empirical content, although they are often thought to have a deep inner significance that is apparent to folks with sensibilities less coarse than those enjoyed by the common run of scientists. Quite the contrary: Myths are for the most part buncombe[2] and cannot be shown not to be so. Thus it is bunk (to choose an example of Lévi-Strauss's) that the touch of a woodpecker's beak can cure toothache, even if there is a deeper inner congruence between tooth and beak that Lévi-Strauss omitted to disclose. It is bunk also that eclipses of the sun come about because the two ravening wolves Hati and Sköll, which hunt the sun daily across the sky, occasionally take a bite out of it. These mythical explanations have empirical explanatory pretensions which are empirically false. What myths and science have in common, as François Jacob has pointed out, is that both are imaginative fabrications—the difference being that myths fail the cruel examination which measures against real life that which purports to be an explanation of it. Bunk has its uses, though: It is fun sometimes to be bunkrapt[3].

Metaphysics has its uses too (under this head I include, for example, the notion of a vital force, of "protoplasm" and of final causes—the idea that the purpose that an organ or a behavioral episode fulfills can have exercised a sort of causal traction which caused it to come into being (as Aristotelian teleology envisages). I include also a great deal of that strange philosophic pastime *Naturphilosophie*. Metaphysics is not nonsense and it is not bunk, for it can be and has been a source of scientific inspiration and of fruitful scientific ideas. It can be, though, a *fraud* and is so wherever it has the explanatory pretensions and address of a scientific explanation. Nearly all of it lies on the wrong side of the track that distinguishes the world of science and common sense from the world of fancy, fiction and metaphysics.

My listeners do not appear to have had any difficulty in understanding what I mean by religion and religious explanations, but if I had been asked what I meant, I should have said I meant explanations that appeal to the agency of a *personal* God—and I should have insisted upon the notion of a personal God, for nothing could be more wishy-washy or generally unsatisfying than to liken God to some kind of diffuse benevolence that permeates the material world.

When I come to formulate the demarcation rules appropriate to the acceptance of the explanatory structures of the transcendent kind, it will be seen that insofar as there is competition between them, religious explanations are by far the best, even though many scientists and philosophers—certainly not all— feel that their acceptance rests upon a kind of apostasy. Certainly their acceptance turns upon faith resting on *belief*— turns upon, as Kant had it, "a kind of consciously imperfect assent."

Belief, then, and faith insofar as it does entail a consciously imperfect assent, entails also a momentary abdication from the rule of reason, something that people who are rationalists by profession are reluctant to accept. This is a very grave mistake on their part—if it is a mistake. If it is not, then it reveals in them a certain kind—their own kind—of fortitude and commitment to principle.

6   The Purpose of Transcendent Explanation and Whether Religion Fulfills It

If, as I gravely fear, the questions about the origin, destiny and purpose of man that Karl Popper called the "ultimate questions" are unanswerable, it is natural to ask the purpose of putting them and of engaging in that which Kant called the "restless endeavor" to answer them—and if we do try to propound answers, in what way shall we distinguish serious answers from those that are merely superficial or whimsical? Clearly there is no point in asking whether they are true or not, at all events if we use truth in the conventional sense of correspondence with reality, because we have already agreed that answers of that kind will not do. We ask these questions because we are in the habit of asking, so deeply has science and commonsense experience inculcated into us the expectation that questions have answers, that there is a reason for all things.

What, then, do we expect of transcendent answers?

I think that scientists, especially, are childlike in the anxiety and spiritual unease aroused in them by incomprehension. When children feel such an anxiety and pester their mothers with "Why" questions, a mother's answers are palliative rather than explanatory. It is not necessary that they be right or even comprehensible—and often they are not. But they give satisfaction enough to bring the exploratory ritual temporarily to a standstill. We, too, whether scientists or laymen, seek peace of mind. The answers we want to hear are those that allay the anxiety of incomprehension and dispel the fear of the dark such as children often have three or four years after the beginning of life and older people may have about the same distance from its end.

Although I believe that the acceptability of transcendent answers must be valued by the degree to which they bring peace of mind, I believe I was mistaken in thinking that empirical congruence—that is, the correspondence of explanation with real life which is the distinguishing mark of scientific explanations—can be left altogether out of account, for whatever else we may expect of transcendent answers, we also expect that they should not be outrageously incongruent with the world of experience and common sense—for if the incongruence is flagrant and barefaced, we shall lose peace of mind. Nowhere is this incongruence more apparent than in the problem of evil and of reconciling the idea of a benevolent God with the natural dispositions and events that are so difficult to reconcile with it. The queen of the sciences has of course grappled with this problem, but not convincingly enough to set our minds at rest.

7   The Question of the Existence of God

Because of the especially important place he occupies in the philosophy of science, I gave Francis Bacon his say near the beginning of this essay. Now, at about the same distance from the end, I think he should be allowed to speak again. Francis Bacon was a simply reverent man, in spite of the traits in his thought that led Paolo Rossi to describe him as "a medieval philosopher haunted by a modern dream."[1] In his Confession of Faith, Bacon wrote thus: "I believe that nothing is without beginning but God; and no nature, no matter, no spirit, but one only and the same God, that God as He is eternally almighty, only wise, only good in His nature, and so He is eternally Father, Son and Spirit in persons."

As the result of some spiritual blindness or deficiency disease, I do not share Bacon's simple reverence, though I know that his belief in God is very widely shared. On the contrary, I believe that a reasonable case can be made for saying, not that we believe in God because He exists but rather that He exists because we believe in Him. In spite of the suspicion that rightly attaches to epigrammatic declarations that are tainted by smartness, the element of truth in the argument I propose to propound has long been recognized in such familiar and flippant blasphemies as "Man created God in his own image."

God and Popper's Third World

In *Pluto's Republic,* I summarized and explained Karl Popper's conception of a third world, inhabited by the creations of the mind, in the following terms:

> Human beings, Popper says, inhabit or interact with three quite distinct worlds: World 1 is the ordinary physical world, or world of physical states; World 2 is the mental world, or world of mental states; the "third world" (you can see why he now prefers to call it World 3) is the world of actual or possible objects of thought—the world of concepts, ideas, theories, theorems, arguments and explanations—the world, let us say, of all the furniture of mind. The elements of this world interact with each other much as the ordinary objects of the material world do: two theories interact and lead to the formulation of a third; Wagner's music influenced Strauss's and his in turn all music written since. Again, I mention that we speak of things of the mind in a revealingly objective way: we "see" an argument, "grasp" an idea, and "handle" numbers expertly or inexpertly as the case may be. The existence of World 3, inseparably bound up with human language, is the most distinctively human of all our possessions. This third world is not a fiction, Popper insists, but exists "in reality." It is a product of the human mind but yet is in large measure autonomous.

In his own account of his idea (*Objective Knowledge: An Evolutionary Approach,* Oxford, 1972), Popper gives more emphasis than I do to the third world's containing theoretical systems, arguments and problem situations.

Considered as an element of this third world, God has the same degree and kind of objective reality as do the other products of mind. It goes with believing in God that we address Him with praise and reverence and obey Him or are otherwise influenced by Him; we make images of Him and believe ourselves to be made in His image. In prayer, we enter into imaginary dialogue with Him and seek from Him comfort and advice. Finally, we believe in God as an agent—indeed, as Prime Mover. God's objective existence rests upon our belief in Him; if that belief were to cease, the reverence and the dialogue would end and we should no longer look to Him as Prime Mover.

Where so many people I like and admire do so and derive strength and comfort from doing so, I am not at all proud of my lack of belief, and while it would not be in my power to simulate belief (a deception that would soon be unmasked), I should like my behavior—short of overt acts of worship or the avowal of beliefs I do not hold—to be such that people take me for a religious man in respect of helpfulness, considerateness and other evidences of an inclination to make the world work better than it otherwise would be. In short, I should like to be thought to possess what it rightly enrages Jewish people to hear described as the "Christian virtues."

I regret my disbelief in God and religious answers generally, for I believe it would give satisfaction and comfort to many in need of it if it were possible to discover and propound good scientific and philosophic reasons to believe in God.

It would not be just to attribute my disbelief to my having led a sheltered academic life less exigently in need of comfort and support than those whose lives have been turbulent or unhappy or in other ways more at risk than my own. Twice in my life I very nearly died as a result of cerebral vascular accidents, and I don't look forward a bit to making, in due course, a clean job of it. I neither cursed God for depriving me of the use of two limbs nor thanked and praised Him for sparing me the use of two others. On these two occasions I derived no comfort from religion or from the thought that God was looking after me; indeed, if I had not disapproved of his famous poem on literary grounds—this kind of braggadocio is a pain—I should have derived more comfort from the William Ernest Henley, who professed to be master of his fate and thanked God for his unconquerable soul. But there was no comfort here. No one knew better than I did that no one is undefeatable: That's just heroics. What matters is not to be defeated. I do not regard myself as either a victim or a beneficiary of divine dispensations, and I do not believe—much though I should like to do so—that God watches over the welfare of small children in the way that small children need looking after (that is, as fond parents do, and pediatricians and good schoolteachers). I do not believe that God does so because there is no reason to believe it. I suppose that's just my trouble: always wanting reasons.

To abdicate from the rule of reason and substitute for it an authentication of belief by the intentness and degree of conviction with which we hold it can be perilous and destructive. Religious belief gives a spurious spiritual dimension to tribal enmities, as we see them in the Low Countries, Ceylon, Northern Ireland and parts of Africa; nor has any religious belief been held with greater passion or degree of conviction than the metaphysics of blood and soil which did so much to animate Hitler's Germany. Was that not also a consequence of just such a deep, passionate conviction as that which has been thought to authenticate religious belief?

The problem of pain has not been solved, though it has been almost hidden from view by a cloud of theological humbug and the still greater exertions of doublethink that conceal from view or pretend the nonexistence of the most unwelcome truth of all. It goes with the passionate intensity and deep conviction of the truth of a religious belief, and of course of the importance of the superstitious observances that go with it, that we should want others to share it—and the only certain way to cause a religious belief to be held by everyone is to liquidate nonbelievers. The price in blood and tears that mankind generally has had to pay for the comfort and spiritual refreshment that religion has brought to a few has been too great to justify our entrusting moral accountancy to religious belief. By "moral accountancy" I mean the judgment that such and such an action is right or wrong, or such a man good and such another evil.

I am a rationalist—something of a period piece nowadays, I admit—but I am usually reluctant to declare myself to be so because of the widespread misunderstanding or neglect of the distinction that must always be drawn in philosophic discussion between the *sufficient* and the *necessary*. I do not believe—indeed, I deem it a comic blunder to believe—that the exercise of reason is *sufficient* to explain our condition and where necessary to remedy it, but I do believe that the exercise of reason is at all times unconditionally *necessary* and that we disregard it at our peril. I and my kind believe that the world can be made a better place to live in (see pp. 39–40)—believe, indeed, that it has already been made so by an endeavor in which, in spite of shortcomings which I do not conceal, natural science has played an important part, of which my fellow scientists and I are immensely proud. I fear that we may never be able to answer those questions about first and last things that have been the subject of this short essay—questions to do with the origin, purpose and destiny of man; we know, however, that whether as individuals or as political people, we do have some say in what comes next, so what could our destiny be except what we make it?

To people of sanguine temperament, the thought that this is so is a source of strength and the energizing force of a just and honorable ambition.

The dismay that may be aroused by our inability to answer questions about first and last things is something for which ordinary people have long since worked out for themselves Voltaire's remedy: "We must cultivate our garden."

Notes

1: *Plus Ultra?*

1. P. B. Medawar, *Pluto's Republic* (Oxford, 1982), p. 326.
2. *Journal of the Warburg and Courtauld Institute*, 1931, 34: 204–17.
3. A professional philosopher chided me for saying this. "Unless it is successful," he told me, "you don't call it science." What rot! I have been engaged in scientific research for about fifty years and I rate it highly scientific even though very many of my hypotheses have turned out mistaken or incomplete. This is our common lot. It is a layman's illusion that in science we caper from pinnacle to pinnacle of achievement and that we exercise a Method which preserves us from error. Indeed we do not; our way of going about things takes it for granted that we guess less often right than wrong, but at the same time ensures that we need not persist in error if we earnestly and honestly endeavor not to do so.
4. Boswell-like, I once asked Karl Popper to express in a sentence the quintessence of the teaching of positivism. He at once replied: "The world is all surface." I had myself until then used a parody of the first proposition of Wittgenstein's *Tractatus logico-philosophicus* (trans. B. A. W. Russell, London, 1922): "The world is everything that seems to be the case"; but joking apart, propositions 1.0 to 2.103 in Wittgenstein are indeed a crash course in positivism.

2: Is the Growth of Science Self-Limited?

1. It sets a lower limit too: Weighing only 2 grams when fully grown, the Etruscan pigmy shrew has so great a surface area in relation to its volume that it is about as small as a free-living warm-blooded animal could be, having regard to the rate of loss of heat, It must in any event eat almost uninterruptedly to maintain its body temperature.
2. To bring the legend of Prometheus fully up to date, we should have it in mind that the heat of the sun is produced by the fusion of hydrogen nuclei to form helium, with a generation of energy proportional to the consequent loss of mass. It was from the sun, then, that the new Prometheus stole the secret of the hydrogen bomb.
3. I use "exosomatic" to refer to our extracorporeal organs, as dialysis machines are exosomatic kidneys and the ventilators used in intensive care are exosomatic lungs. Computers are essentially exosomatic brains, for many of their functions are brainlike.

3: Is There an Intrinsic Limitation upon the Growth of Science?

1. Who would magnify a newspaper photograph—composed, as it is, of millions of tiny little dots—in order to see more detail in it?
2. A nanometer is one billionth of a meter (10^{-9}m).

4: Where *Plus Ultra* Prevails

1. Immanuel Kant would have been unwilling to believe the ultimate questions unanswerable. If that had been so, he asks in the Preface to the second edition of *Critique of Pure Reason*, why should nature have visited our reason with the restless endeavor whereby it is ever searching for answers, as if this were one of its most important concerns? (Norman Kemp Smith's translation, Oxford, 1929, p. 21). However, Dr. Samuel

Johnson, so often our point of attachment to common sense, clearly believed that some questions *were* unanswerable—a judgment of considerable importance because he was a deeply thoughtful and reverent man. Boswell cites one such example: Johnson had said, "There are innumerable questions to which the inquisitive mind in this state can conceive no answer— Why do you and I exist? Why was the world created? Since it was to be created, why was it not created sooner?"

2. Anyone who believes that the imagery of *The Rime of the Ancient Mariner* or of *Kubla Khan* arose de novo in Coleridge's mind will have his illusions dispelled by John Livingston Lowes's *The Road to Xanadu* (1927).

3. *The Structure of Scientific Revolutions* (University of Chicago Press, 1962; 1970); *Essential Tension* (University of Chicago Press, 1978).

5: Where Then Shall We Turn?

1. I used here, as most people would incline to, "transcendental," but Karl Popper told me that Kant himself (whose opinion here is law-giving) would have used "transcendent," meaning outside the domain of actual or possible empirical sense experience and natural science or common sense.

2. Buncombe, *abbr.* "bunk": "A kind of vaporous talk associated especially with the congressional member for Buncombe, North Carolina." —C. T. Onions, *Oxford Dictionary of English Etymology* (1966).

3. A word inadvertently coined by Paul Jennings when he'd intended to type "bankrupt."

7: The Question of the Existence of God

1. *Francis Bacon: From Magic to Science*, trans. S. Rabinovitch (London: Routledge & Kegan Paul, 1968).

Index

OXFORD

MORE OXFORD PAPERBACKS

Details of a selection of other books follow. A complete list of Oxford Paperbacks, including The World's Classics, Twentieth-Century Classics, OPUS, Past Masters, Oxford Authors, Oxford Shakespeare, and Oxford Paperback Reference, is available in the UK from the General Publicity Department, Oxford University Press (JH), Walton Street, Oxford, OX2 6DP.

In the USA, complete lists are available from the Paperbacks Marketing Manager, Oxford University Press, 200 Madison Avenue, New York, NY 10016.

Oxford Paperbacks are available from all good bookshops. In case of difficulty, please order direct from Oxford University Press Bookshop, 116 High Street, Oxford, Freepost, OX1 4BR, enclosing full payment. Please add 10% of published price for postage and packing.

MARTIN CHUZZLEWIT

Charles Dickens

Edited by Margaret Cardwell

Martin Chuzzlewit—hailed as the most ambitious of Dickens's early novels—has frequently been likened to Franny Trollope's book, *The Domestic Manners of the Americans,* for the criticism it aroused by its controversial opinions of the Americans.

This edition, with a new introduction and explanatory notes, includes Dickens's Prefaces and the 1868 Postscript which gave his later views of America, the few working notes which accompany the manuscript, and eight of the original illustrations by 'Phiz'.

The World's Classics

OUR VILLAGE

Mary Russell Mitford

Illustrated by Joan Hassall
Introduction by Margaret Lane

'Of all the situations for a constant residence, that which appears to me most delightful is a little village, far in the country . . . with inhabitants whose faces are as familiar to us as flowers in our garden.' Mary Russell Mitford lived in just such a village, Three Mile Cross in Berkshire, for more than thirty years. She drew on her observations of the locality for many short essays, the best of which appear in this book, which give a unique picture of country life in the early years of the nineteenth century.

ENGLISH HOURS

Henry James

Introduction by Leon Edel

'The great relisher of impressions and nuances turns his attention in these occasional pieces to aspects of English life, metropolitan and provincial, Wonderful.' *Sunday Times*

English Hours is an affectionate portrait, warts and all, of the country that was to become Henry James's adopted homeland. One of the great travel writers of his time, James takles the reader on a series of visits, ranging from Winchelsea to Warwick, taking in abbeys and castles, sea-fronts and race-courses. Though no mere travel guide, the book will certainly enhance any tourist's pleasure.

COTTAGE ECONOMY

William Cobbett

With a preface by G. K. Chesterton

'I view the tea drinking as a destroyer of health, an enfeebler of the frame, an engenderer of effeminacy and laziness, a debaucher of youth and a maker of misery for old age.' First published in 1822, Cobbett's classic handbook for smallholders was many times revised and enlarged, and is now reissued in its latest edition (1850). Cobbett tells us, among much else, how to brew beer, make bread, keep cows, pigs, bees, ewes, goats, poultry, and rabbits. And the book is full of splendid passages of social and political invective, making it a manifesto for Cobbett's philosophy of self-sufficiency. 'A couple of flitches of bacon are worth fifty thousand Methodist sermons and religious tracts.'

'a masterpiece of English dottiness' *The Times*

A LITTLE TOUR IN FRANCE

Henry James

Foreword by Geoffrey Grigson

One rainy morning in the autumn of 1882 Henry James set out on a trip to the French provinces, which took him from Touraine to the south-west, through Provence, and northwards again along the flooding Rhone to Burgundy.

James is a knowledgeable guide. His verbal discussions make no pedantic tourists' handbook; and if, as Geoffrey Grigson points out, they lead the reader round not only France but also the country called Henry James, lovers of both countries, will find that each in this book exerts an irresistible charm.

ADVICE TO YOUNG MEN

William Cobbett

Cobbett's accounts in this book of learning grammar in an army barracks, selecting a girl to marry (and nearly betraying his chosen bride), warding off barking dogs on a hot night in Philadelphia, and many other equally diverse activities, display to the full the talents which brought him the widest readership of any journalist of his day on either side of the Atlantic. The six 'Letters' that make up *Advice to Young Men* are his prescription for those who wish to be happy, addressed to a Youth, a Bachelor, a Lover, a Husband, a Father, and a Citizen.

'Even when he's talking nonsense about the bad effect of the reading of "romances" . . . or of "the punning and smutty Shakespeare", he writes with a sturdy vividness that must give pleasure to anyone with an ear.' *Times Educational Supplement*

'Good, austere, wholesome, sonorously argued precepts on the surface, it would take a great deal of courage to carry them out on the domestic front, then or now.' Alex Hamilton, *Guardian*

TRAVELS THROUGH FRANCE AND ITALY

Tobias Smollett

Edited by Frank Felsenstein

Whether he describes the French tax system, the culture of silkworms, or the marbles of Florence, Smollett provides many insights into eighteenth-century taste and into his own cantankerous, perceptive and intelligent personality.

'any reader of this new edition of his *Travels* . . . will be amused, horrified and aghast—aghast at Smollett (sometimes) as well as at the French of the eighteenth century (some of them).' Geoffrey Grigson, *Country Life*.

The World's Classics

THE HANGMAN'S DAUGHTER AND OTHER STORIES

Daisy Ashford

Introduced by Margaret Steel

Daisy Ashford's stories about love, marriage, and the social foibles of the Victorian world have delighted many generations of readers. Ashford family tradition has it that Daisy, who was born in 1881, composed her first tale, 'The Life of Father McSwiney', when she was just four. Only recently rediscovered, and first published in Oxford Paperbacks, it tells of a Jesuit priest whose astonishing adventures include a sightseeing visit to London with the Pope. The collection also includes two other stories, 'Where Love Lies Deepest' and 'The Hangman's Daughter,' and will not fail to delight those who have enjoyed *The Young Visiters* and *Love and Marriage*. Daisy Ashford's elder daughter has contributed a charming introduction describing the family that nurtured her mother's precocious literary talent.

LAVENGRO

George Borrow

Now acknowledged as a classic, *Lavengro* is a book in which autobiography is inextricably linked with fiction. Borrow was an inveterate traveller with a taste for the outlandish. The restless spirit of the young hero of *Lavengro* leads him to strange lands and adventures with gypsies, rogues and thieves.

'One of the more remarkable literary oddities of the nineteenth century: an autobiographical fantasy journey.' *Observer*

THE ROMANY RYE

George Borrow

The sequel to *Lavengro* which continues Borrow's adventurous tale of gypsy life.

HAMPSHIRE DAYS

W. H. Hudson

First published in 1903, this is a personal celebration of the delights of Hampshire—its abundant wildlife, mysterious barrows, impressive New Forest, and its human 'characters'. Hudson shares with Gilbert White the gift of making his observations in a vividly readable style. The result is a sensuous kaleidoscope of the colours, scents, and songs of the English countryside.

'A classic volume of natural history written in 1903; whether noting the behaviour of a baby cuckoo in a robin's nest or commenting on the mutual mistrust of dark-eyed and fair country people, Hudson's observations of a nearly vanished world are superb.' *Sunday Times*